T0140144

Springer Theses

Recognizing Outstanding Ph.D. Research

Aims and Scope

The series "Springer Theses" brings together a selection of the very best Ph.D. theses from around the world and across the physical sciences. Nominated and endorsed by two recognized specialists, each published volume has been selected for its scientific excellence and the high impact of its contents for the pertinent field of research. For greater accessibility to non-specialists, the published versions include an extended introduction, as well as a foreword by the student's supervisor explaining the special relevance of the work for the field. As a whole, the series will provide a valuable resource both for newcomers to the research fields described, and for other scientists seeking detailed background information on special questions. Finally, it provides an accredited documentation of the valuable contributions made by today's younger generation of scientists.

Theses are accepted into the series by invited nomination only and must fulfill all of the following criteria

- They must be written in good English.
- The topic should fall within the confines of Chemistry, Physics, Earth Sciences, Engineering and related interdisciplinary fields such as Materials, Nanoscience, Chemical Engineering, Complex Systems and Biophysics.
- The work reported in the thesis must represent a significant scientific advance.
- If the thesis includes previously published material, permission to reproduce this must be gained from the respective copyright holder.
- They must have been examined and passed during the 12 months prior to nomination.
- Each thesis should include a foreword by the supervisor outlining the significance of its content.
- The theses should have a clearly defined structure including an introduction accessible to scientists not expert in that particular field.

More information about this series at http://www.springer.com/series/8790

Daniel Salerno

The Higgs Boson Produced With Top Quarks in Fully Hadronic Signatures

Doctoral Thesis accepted by
the University of Zurich, Zürich, Switzerland

Author
Dr. Daniel Salerno
Department of Physics
University of Zurich
Zürich, Switzerland

Supervisor
Prof. Dr. Florencia Caneli
Department of Physics
University of Zurich
Zürich, Switzerland

ISSN 2190-5053 ISSN 2190-5061 (electronic)
Springer Theses
ISBN 978-3-030-31259-6 ISBN 978-3-030-31257-2 (eBook)
https://doi.org/10.1007/978-3-030-31257-2

This Springer imprint is published by the registered company Springer Nature Switzerland AG
The registered company address is: Gewerbestrasse 11, 6330 Cham, Switzerland

Supervisor's Foreword

The discovery of the Higgs boson by the CMS and ATLAS collaborations at the Large Hadron Collider (LHC) has opened up the way for exploring its interactions with other particles. Thanks to an unprecedented luminosity at the LHC, enough Higgs bosons have been produced, and all of its main production and decay modes have now been observed. Nonetheless, some properties of the Higgs boson have not yet been established. The main goal of this Ph.D. thesis has been the elucidation of one of these properties, the tree-level coupling of Higgs bosons to top quarks.

In the Standard Model, the Higgs boson couples to fermions in a manner described by the Yukawa sector, with a coupling strength proportional to the mass of the fermions. For the heaviest known fundamental fermion, the top quark, the Yukawa coupling can be measured in a model-independent way in the associated production of a Higgs boson and a top quark-antiquark ($t\bar{t}$) pair, in a process called "$t\bar{t}H$ production". This measurement is usually pursued in the different decay channels of the Higgs boson and of the $t\bar{t}$ pair. In particular, the fully hadronic channel, is the main topic of Daniel Salerno's Ph.D. thesis. In this signature the Higgs boson decays into a bottom quark-antiquark pair and the top quark pair decays only to jets, leaving a total of four jets which contain the decay products of B hadrons (b jets). Even though this is the most common decay signature for the $t\bar{t}H$ process, the large number of jets from the decays makes the process challenging to unequivocally reconstruct and the huge contamination of "irreducible" backgrounds with signatures very similar to the $t\bar{t}H$ signal makes it difficult to extract from the data.

In Chap. 1 the author provides a summary of the Standard Model (SM) of particle physics and the theoretical background which includes a review of the Higgs mechanism and the phenomenology of the Higgs boson at the LHC, as well as presenting an up-to-date summary of recent measurements of Higgs boson production and properties. A description of the LHC accelerator and CMS experiment is given in Chap. 2. In Chap. 3 the reconstruction of the proton-proton collisions recorded by the CMS experiment is described, with emphasis on the final-state physics objects used in the analysis. To enrich the $t\bar{t}H$ signal, the author developed dedicated trigger algorithms essential in the extraction of the fully-hadronic $t\bar{t}H$ events from the proton-proton collision data. Details of this work are also described in Chap. 3. Chapter 4 is dedicated to describing the implementation of the matrix-element approach to aid discrimination between $t\bar{t}H$ and SM background events. This is a crucial component of the thesis as the large number of jets in the signature leads to a huge number of possible combinations for specifying $t\bar{t}H$ and thus an impractical computational time. Chapter 5 concentrates on the analysis strategy from data collection and the Monte Carlo (MC) samples used to the calculation of the systematic uncertainties. In particular, Chap. 5 describes the implementation of a new technique to estimate and reduce the strongly produced QCD multijet background based on the quark-gluon discrimination of jets, which enables differentiation between gluon-rich background events, and quark-rich $t\bar{t}H$ signal events. The final results are presented in Chap. 6 after a detailed explanation of the statistical treatment used to extract the results. This chapter also includes a comparison with previous results by CMS in complementary channels and by the ATLAS experiment. Chapter 7 is reserved for conclusions. The final Chap. 8 is an outlook summarizing the possible improvements and sensitivity reach of the analysis for future LHC datasets.

The topics studied in Daniel Salerno's thesis cover a vast array of investigations and creative developments. They are described in a detailed, complete, and accessible manner and I believe this work will be useful for beginners and experts alike in the area of fundamental science.

Zürich, Switzerland Prof. Dr. Florencia Canelli
July 2019

Abstract

I present my work at the CMS experiment on a search for the standard model (SM) Higgs boson produced in association with top quarks. The search is targeted towards final states compatible with the $H \rightarrow b\bar{b}$ decay and the fully hadronic decay channel of the $t\bar{t}$ pair, and uses data from proton-proton collisions at a centre-of-mass energy of 13 TeV, corresponding to an integrated luminosity of 35.9 fb^{-1}. This is a challenging search with many final state particles that cannot be uniquely identified and with large contamination from SM background processes. It is performed for the first time at CMS and the first time anywhere at $\sqrt{s} = 13\,\text{TeV}$, and contributes to the overall sensitivity of the $t\bar{t}H$ cross-sectional measurement, which constitutes a crucial test of the SM.

The CMS apparatus is a multipurpose detector operating at the LHC, which is a hadron collider at CERN. The CMS detector operates a 3.8 T superconducting solenoid, and includes dedicated subsystems for charged particle tracking near the interaction point, measurements of electromagnetic and hadronic energy deposits, and muon tracking outside the solenoid, all of which provide nearly 4π coverage. The trigger and data acquisition system of CMS, to which I made original contributions, efficiently reduces the event rate from the 40 MHz collision rate to around 1 kHz for permanent storage and offline analysis. I also developed dedicated jet based triggers for the fully hadronic $t\bar{t}H$ search.

I performed all aspects of the search, making original contributions to all techniques and measurements specific to it. A jet based quark-gluon discriminator is used in an event-based likelihood ratio for the first time in a CMS search to differentiate between events containing jets originating from light-flavour quarks and events containing jets from gluons. A unique method to estimate the dominant QCD multijet background from data is developed. Selected events with 7 or more jets and 3 or more b-tagged jets are allocated to one of 6 categories based on jet and b-tag multiplicity, with different levels of signal purity.

A matrix element method (MEM) is used for optimal discrimination between the $t\bar{t}H$ signal and SM background processes and for the ultimate signal extraction. It assigns a signal and background probability density to each event using the full event information and leading order matrix amplitudes of the $t\bar{t}H$ and $t\bar{t}+b\bar{b}$ processes. It sums over all combinations of jet-quark associations to reduce the uncertainty of matching the correct pairs, and it integrates over poorly measured or missing variables. A likelihood ratio of these two probability densities is used to form the final MEM discriminant. This is the first time that the MEM has been used in a fully hadronic final state at CMS.

The results are interpreted via an observed $t\bar{t}H$ signal strength relative to the SM cross section under the assumption of $m_{\mathrm{H}} = 125\,\mathrm{GeV}$, i.e. $\mu = \sigma/\sigma_{\mathrm{SM}}$. A binned maximum likelihood fit is performed to the MEM discriminant in all categories to extract a best-fit value of $\hat{\mu} = 0.9 \pm 1.5$. This is compatible with the SM prediction of $\mu = 1$, and corresponds to observed and expected significances of 0.6 and 0.7 standard deviations, respectively. Under the background-only hypothesis, upper exclusion limits on the signal strength of $\mu < 3.8$ and $\mu < 3.1$ are observed and expected, respectively, at the 95% confidence level.

Acknowledgements

First and foremost, I would like to express my deepest gratitude to my supervisor, Prof. Dr. Florencia Canelli, for giving me the opportunity to work on such an impressive search within a highly esteemed collaboration at the forefront of particle physics research. Her guidance has been instrumental in getting me to this point, in developing my skills as a physicist, and in bringing this complex search to completion.

I would also like to express my sincere gratitude and appreciation to Dr. Lea Caminada who has supported me throughout this endeavour with all aspects of the search and physics research in general.

I would like to thank the members and former members of ETH, with whom I shared a close collaboration for the duration of my studies. In particular, my deepest thanks to Dr. Lorenzo Bianchini for not only introducing me to data analysis at CMS, but also for his development of the matrix element algorithm used in the search. I would also like to express my gratitude to Joosep Pata, for his development of the framework used in the search and his support in all technical aspects of the search.

I would like to thank my university colleagues, especially Dr. Silvio Donato for his guidance and support and Korbinian Schweiger for his contribution to the search. In addition, a special thank you to my office mates Deborah Pinna and Camilla Galloni, who created the most enjoyable working environment during my studies.

Last but by no means least, I would to like express my deepest gratitude to Valentina Zingg for her unconditional love and support throughout my studies.

Contents

Chapter 1
Introduction

The standard model of particle physics represents one of the great successes of elementary particle physics in recent times, being able to predict various physics processes and observables that have later been experimentally confirmed. The final element to be verified through experiment is the presence of a Higgs boson, the particle associated with the field that generates the mass of all elementary particles. In 2012, the ATLAS and CMS experiments at the LHC observed a new boson with a mass of approximately 125 GeV [1, 2]. Many measurements of the properties of this new boson have been performed to date, and all have been found to be consistent with predictions from the standard model [3–6], indicating that the new particle is indeed the standard model Higgs boson.

This discovery not only represents a great success of the standard model but also a great achievement for the LHC, a hadron accelerator and collider operating at CERN. Since 2010, the LHC has provided proton–proton collisions at centre-of-mass energies of 7, 8 and 13 TeV. It was primarily built to find the Higgs boson, but it also offers great prospects to search for many new physics processes and particles predicted by modern theories beyond the standard model. The CMS experiment is one of two general purpose experiments at the LHC designed to probe high energy and high intensity proton–proton collisions for signs of new physics and, in particular, the Higgs boson. The discovery of the Higgs boson is a testament to the outstanding performance of the CMS detector, which is able to measure the energy and position of particles produced in proton–proton collisions with extreme precision.

Between CMS and ATLAS, the Higgs boson has been observed in almost all decays modes, namely the $\gamma\gamma$ [7, 8], ZZ [9, 10], W^+W^- [11, 12] and $\tau^+\tau^-$ [13] final states, while strong evidence has been reported in the $b\bar{b}$ final state [14, 15]. These observations all confirm the standard model couplings of the Higgs boson to vector bosons, τ leptons and bottom quarks. The interaction of the Higgs boson with fermions is governed by the Yukawa interaction, and its strength is given by the Yukawa coupling, which is proportional to the fermion mass. An important Yukawa

D. Salerno, *The Higgs Boson Produced With Top Quarks in Fully Hadronic Signatures*, Springer Theses, https://doi.org/10.1007/978-3-030-31257-2_1

coupling is that of the Higgs boson to the top quark, which has been measured indirectly in loop processes involving top quarks, i.e. in Higgs boson production through gluon-gluon fusion and Higgs boson decays to photons. However, a direct measurement of the top-Higgs coupling is essential to avoid potential influences from beyond standard model processes, which can enter in the loop unnoticed. In this regard, the search for the Higgs boson produced in association with top quarks ($\mathrm{t\bar{t}H}$ production) is crucial to our understanding of the standard model.

A direct measurement of the top quark Yukawa coupling would also be the first measurement of the Higgs boson coupling to an up-type fermion, providing a direct inspection of possible inequivalent Yukawa couplings of up- and down-type fermions—a disfavoured non-standard model coupling hypothesis that has thus far only been measured indirectly. In addition to a direct measurement of the top-Higgs coupling, $\mathrm{t\bar{t}H}$ production can provide insights to beyond standard model physics. Given the relatively large top quark mass ($m_{\mathrm{t}} \approx 172.5\,\mathrm{GeV}$), and the importance of the role the Higgs boson plays in providing mass to the fundamental particles, deviations in the observed $\mathrm{t\bar{t}H}$ production cross section from the predictions of the standard model can indicate the presence of yet-unseen dynamics in the electroweak sector.

The CMS experiment has already performed a number of searches for $\mathrm{t\bar{t}H}$ production using 7 and 8 TeV collision data from 2011 and 2012, corresponding to $5\,\mathrm{fb}^{-1}$ and $19.5\,\mathrm{fb}^{-1}$, respectively [16, 17]. New results at a centre-of-mass energy of 13 TeV have been obtained in the $\mathrm{W^+W^-}$/multiple-lepton [18], ZZ [19], $\gamma\gamma$ [20] and $\tau^+\tau^-$ [21] final-states of the Higgs boson with $35.9\,\mathrm{fb}^{-1}$ of data collected in 2016. The ATLAS experiment has performed similar $\mathrm{t\bar{t}H}$ searches and has provided evidence of $\mathrm{t\bar{t}H}$ production by combining final states [22]. With the results of this thesis and other $\mathrm{t\bar{t}H}$ analyses, CMS has performed a combination and reported the first ever observation of $\mathrm{t\bar{t}H}$ production [23].

A particularly important subprocess of $\mathrm{t\bar{t}H}$ production is when then Higgs boson subsequently decays to bottom quarks. It is unique due to a very specific Higgs coupling space: all couplings are fermionic and restricted to the third-generation quarks only. As a consequence, the results obtained in the $\mathrm{H \to b\bar{b}}$ decay channel should be easier to interpret than those in other decay modes. At CMS, a first search at $\sqrt{s} = 13\,\mathrm{TeV}$ for $\mathrm{t\bar{t}H}$ production in the $\mathrm{H \to b\bar{b}}$ final state in which at least one top quark decays leptonically was conducted with $2.7\,\mathrm{fb}^{-1}$ of data collected in 2015 [24]. This search was later extended using the first $12.9\,\mathrm{fb}^{-1}$ of data collected in 2016 [25], and most recently presented with the full 2016 data set [26]. The ATLAS experiment has released results for the $\mathrm{t\bar{t}H}$ search in the $\mathrm{H \to b\bar{b}}$ final state using $36.1\,\mathrm{fb}^{-1}$ of 13 TeV data [27].

The focus of this thesis is the search for $\mathrm{t\bar{t}H}$ production in the fully hadronic decay channel, where the Higgs boson decays exclusively to a bottom quark-antiquark pair, and each top quark decays to a bottom quark and a W boson that decays to two light quarks. ATLAS has already performed a search in the fully hadronic final state of $\mathrm{t\bar{t}H}$ at 8 TeV [28]. At CMS however, the first result in this channel has been performed at 13 TeV [29], and is the subject of this thesis. The analysis uses proton–proton collision data delivered by the LHC and collected by CMS in 2016 at a centre-of-mass energy of 13 TeV, corresponding to an integrated luminosity of $35.9\,\mathrm{fb}^{-1}$. The final state of this process involves eight quarks, four of which are b quarks. Ideally, the signal would therefore appear in the CMS detector as eight jets, of which four are tagged as b jets by a software algorithm. To accommodate jets lost to detector acceptance, merging of separate quarks, and the efficiency of tagging b jets, events with seven or more jets and three or more b jets are analysed. To account for extra jets from initial or final-state radiation, up to nine jets are considered per event.

Although the signature discussed involves a large number of high-p_{T} final-state jets, the absence of leptons essentially ensures it suffers from a very large background contamination. By far, the dominant background is from jets produced through the strong interaction, referred to as QCD multijet events. Further complicating the analysis, is a large contribution from $\mathrm{t\bar{t}}$ + jets production, including $\mathrm{t\bar{t}}$ + light-flavour jets, where one or more of the jets are incorrectly identified as b jets, as well as $\mathrm{t\bar{t}} + \mathrm{c\bar{c}}$, and the irreducible $\mathrm{t\bar{t}} + \mathrm{b\bar{b}}$ background. Smaller background contributions arise from other standard model processes. A technique to reduce the contribution of QCD multijet events, based on the quark-gluon discrimination of jets, has been used for the first time at CMS.

Given the many combinations of jet-quark matching, it is not possible to resolve a clear Higgs boson resonant mass peak. Nevertheless, there are underlying kinematic differences between the $\mathrm{t\bar{t}H}$ signal and the multijet background and, to a lesser extent, the $\mathrm{t\bar{t}}$ + jets background. These differences are exploited through the use of a matrix element method to distinguish signal from background events. Specifically, events are assigned a probability density according to how compatible they are with the lowest "tree" level $\mathrm{t\bar{t}H}$ process. Although this probability density alone provides some separation between the signal and most background processes, a second probability density is assigned to each event according to its compatibility with the tree level $\mathrm{t\bar{t}} + \mathrm{b\bar{b}}$ process, which provides extra discrimination against the irreducible $\mathrm{t\bar{t}} + \mathrm{b\bar{b}}$ background. The two probability densities are combined into a likelihood ratio to form the final discriminant of the analysis.

Chapter 2 of this thesis provides a thorough description of the theoretical framework from which the analysis arises. First a brief overview of the standard model, including historical pretext, is given. Then detailed derivations of the theoretical models which lead to the Higgs boson are reproduced. The properties of the Higgs boson in the standard model, and its interaction with other particles and itself are also discussed. Theoretical calculations of the Higgs boson decay and production rates are given followed by the latest experimental results from the LHC. Finally, a detailed look at the particular production and decay channel of the analysis is presented, along with some information on the standard model background processes which could mimic the signal.

In Chap. 3, a description of the experimental apparatus is provided. The chapter begins with a brief introduction to the LHC, which is the hadron accelerator and collider providing the high-energy proton–proton collisions that are studied. Then a description of the CMS detector follows. It is a general purpose detector with a large superconducting solenoid magnet as its central feature. Within the solenoid reside a silicon tracker for precision measurements of charged particles close to the interaction point, and calorimeters with large forward coverage for measurements of particle energy and missing transverse energy. Outside the solenoid, muon detectors capture the tracks of muons, most of which penetrate the entire detector. The complex trigger and data acquisition systems, to which I made original contributions, form a crucial component of CMS and are responsible for providing the high quality data to be analysed. Initial details about the reconstruction of particles within each subdetector are provided, which are then built upon in Chap. 4.

A detailed description of the trigger requirements and the selection criteria on reconstructed particles is provided in Chap. 4. I developed the all-jet triggers used in the search specifically for this analysis. They consist of a sequence of requirements on the number and transverse momentum of jets, as well as the number of b jets. The particles used in the analysis are reconstructed with the particle-flow algorithm, which is a hallmark of the CMS experiment used in nearly all of its analyses. It exploits the outstanding spatial and energy resolution and almost 4π coverage of the CMS detector to reconstruct all stable particles in an event, combining the information from all subdetectors. It is especially performant in the identification and reconstruction of jets and missing transverse momentum. The identification of b jets is performed by a dedicated "b tagging" algorithm, while an algorithm used to distinguish jets originating from light flavour quarks and gluons is used in the event selection.

Chapter 5 is dedicated to the matrix element method, the technical algorithm used as the final discriminant between signal and background. It introduces the general concept of the method before delving into its details. The theoretical foundation of the algorithm is described along with the simplifying assumptions used in its implementation. The technical aspects are discussed and the construction of the final discriminant is reproduced. Finally, the validation and performance of the method are presented. This is the first time the matrix element method has been used in an all-jet final state at CMS. I made original contributions to the development and implementation of the algorithm for use in this and other $t\bar{t}H$ searches.

In Chap. 6, the analysis strategy is described in detail. Beginning with the description of the data and simulation samples used, I go on to explain the reweighting methods applied to simulation, which are needed to account for differences in simulation modelling with respect to data. The criteria used to select signal events, including their categorisation, are then given followed by a description of the background estimation methods. The signal extraction is explained along with the final event yields and discriminant distributions for the signal and background processes. Finally, the systematic uncertainties affecting the analysis are discussed. I performed all aspects of the analysis myself, with some support from colleagues, and developed all analysis-specific techniques and measurements.

The results of the search are presented in Chap. 7. The statistical method used to extract the signal from the background is described and demonstrated. The results are presented in terms of the signal strength modifier μ, which is defined as the ratio of the measured $t\bar{t}H$ production cross section to the standard model prediction, given a 125 GeV Higgs boson mass. Due to the relatively low significance observed, different interpretations of the signal strength are provided. A 95% confidence level upper limit on μ is given assuming a background-only hypothesis. The best-fit value for μ is calculated as well as the significance of the signal over the background-only hypothesis. The results of a combination with other $t\bar{t}H$ searches at CMS are also presented and the contribution of this analysis to the combined result is discussed.

The thesis concludes with a summary of the methods and results and a brief discussion of the future prospects for the analysis.

References

1. CMS Collaboration (2012) Observation of a new boson at a mass of 125 GeV with the CMS experiment at the LHC. Phys Lett B 716:30. https://doi.org/10.1016/j.physletb.2012.08.021. arXiv:1207.7235
2. ATLAS Collaboration (2012) Observation of a new particle in the search for the Standard Model Higgs boson with the ATLAS detector at the LHC. Phys Lett B 716:1. https://doi.org/10.1016/j.physletb.2012.08.020. arXiv:1207.7214
3. ATLAS Collaboration (2016) Measurements of the Higgs boson production and decay rates and coupling strengths using pp collision data at $\sqrt{s} = 7$ and 8 TeV in the ATLAS experiment. Eur Phys J C 76:6. https://doi.org/10.1140/epjc/s10052-015-3769-y, arXiv:1507.04548
4. CMS Collaboration (2015) Precise determination of the mass of the Higgs boson and tests of compatibility of its couplings with the standard model predictions using proton collisions at 7 and 8 TeV. Eur. Phys. J. C 75:212. https://doi.org/10.1140/epjc/s10052-015-3351-7. arXiv:1412.8662
5. CMS Collaboration (2013) Study of the Mass and Spin-Parity of the Higgs Boson Candidate Via Its Decays to Z Boson Pairs. Phys Rev Lett 110:081803. https://doi.org/10.1103/PhysRevLett.110.081803. arXiv:1212.6639
6. ATLAS Collaboration (2013) Evidence for the spin-0 nature of the Higgs boson using ATLAS data. Phys Lett B 726:120. https://doi.org/10.1016/j.physletb.2013.08.026. arXiv:1307.1432
7. ATLAS Collaboration (2014) Measurement of Higgs boson production in the diphoton decay channel in pp collisions at center-of-mass energies of 7 and 8 TeV with the ATLAS detector. Phys Rev D 90:112015. https://doi.org/10.1103/PhysRevD.90.112015. arXiv:1408.7084
8. CMS Collaboration (2014) Observation of the diphoton decay of the Higgs boson and measurement of its properties. Eur Phys J C 74:3076. https://doi.org/10.1140/epjc/s10052-014-3076-z. arXiv:1407.0558
9. ATLAS Collaboration (2015) Measurements of Higgs boson production and couplings in the four-lepton channel in pp collisions at center-of-mass energies of 7 and 8 TeV with the ATLAS detector. Phys Rev D 91:012006. https://doi.org/10.1103/PhysRevD.91.012006. arXiv:1408.5191
10. CMS Collaboration (2014) Measurement of the properties of a Higgs boson in the four-lepton final state. Phys Rev D 89:092007. https://doi.org/10.1103/PhysRevD.89.092007. arXiv:1312.5353
11. ATLAS Collaboration (2015) Study of (W/Z)H production and Higgs boson couplings using $H \to WW^*$ decays with the ATLAS detector. JHEP 08:137. https://doi.org/10.1007/JHEP08(2015)137, arXiv:1506.06641
12. CMS Collaboration (2014) Measurement of Higgs boson production and properties in the WW decay channel with leptonic final states. JHEP 01:096. https://doi.org/10.1007/JHEP01(2014)096. arXiv:1312.1129
13. CMS Collaboration (2018) Observation of the Higgs boson decay to a pair of τ leptons with the CMS detector. Phys Lett B 779:283. https://doi.org/10.1016/j.physletb.2018.02.004, arXiv:1708.00373

14. ATLAS Collaboration (2017) Evidence for the $H \rightarrow b\bar{b}$ decay with the ATLAS detector. JHEP 12:024. https://doi.org/10.1007/JHEP12(2017)024, arXiv:1708.03299
15. CMS Collaboration (2018) Evidence for the Higgs boson decay to a bottom quark-antiquark pair. Phys Lett B 780:501. https://doi.org/10.1016/j.physletb.2018.02.050, arXiv:1709.07497
16. CMS Collaboration (2014) Search for the associated production of the Higgs boson with a top-quark pair. JHEP 09:087. https://doi.org/10.1007/JHEP10(2014)106, arXiv:1408.1682. [Erratum: JHEP 10:106(2014)]
17. CMS Collaboration (2015) Search for a standard model higgs boson produced in association with a top-quark pair and decaying to bottom quarks using a matrix element method. Eur Phys J C 75:251. https://doi.org/10.1140/epjc/s10052-015-3454-1, arXiv:1502.02485
18. CMS Collaboration (2017) Search for Higgs boson production in association with top quarks in multilepton final states at $\sqrt{s} = 13$ TeV. CMS-PAS-HIG-17-004.https://cds.cern.ch/record/2256103
19. CMS Collaboration (2017) Measurements of properties of the Higgs boson decaying into the four-lepton final state in pp collisions at $\sqrt{s} = 13$ TeV. JHEP 11:047. https://doi.org/10.1007/JHEP11(2017)047, arXiv:1706.09936
20. CMS Collaboration (2018) Measurements of Higgs boson properties in the diphoton decay channel in proton-proton collisions at $\sqrt{s} = 13$ TeV. JHEP 11:185. https://doi.org/10.1007/JHEP11(2018)185, arXiv:1804.02716
21. CMS Collaboration (2017) Search for the associated production of a Higgs boson with a top quark pair in final states with a τ lepton at $\sqrt{s} = 13$ TeV. CMS-PAS-HIG-17-003.https://cds.cern.ch/record/2257067
22. ATLAS Collaboration (2018) Evidence for the associated production of the Higgs boson and a top quark pair with the ATLAS detector. Phys Rev D97(7):072003. https://doi.org/10.1103/PhysRevD.97.072003, arXiv:1712.08891
23. CMS Collaboration (2018) Observation of $\mathrm{t\bar{t}H}$ production. Phys Rev Lett 120(3):231801. https://doi.org/10.1103/PhysRevLett.120.231801, arXiv:1804.02610
24. CMS Collaboration (2016) Search for $\mathrm{t\bar{t}H}$ production in the $\mathrm{H} \rightarrow \mathrm{b\bar{b}}$ decay channel with $\sqrt{s} = 13$ TeV pp collisions at the CMS experiment. CMS-PAS-HIG-16-004.https://cds.cern.ch/record/2139578
25. CMS Collaboration (2016) Search for $\mathrm{t\bar{t}H}$ production in the $\mathrm{H} \rightarrow \mathrm{b\bar{b}}$ decay channel with 2016 pp collision data at $\sqrt{s} = 13$ TeV. CMS-PAS-HIG-16-038.https://cds.cern.ch/record/2231510
26. CMS Collaboration (2019) Search for $\mathrm{t\bar{t}H}$ production in the $\mathrm{H} \rightarrow \mathrm{b\bar{b}}$ decaychannel with leptonic $\mathrm{t\bar{t}}$ decaysin proton-proton collisions at $\sqrt{s} = 13$ TeV. JHEP 03:026.https://doi.org/10.1007/JHEP03(2019)026, arXiv:1804.03682
27. ATLAS Collaboration (2018) Search for the standard model Higgs boson produced in association with top quarks and decaying into a $b\bar{b}$ pair in pp collisions at $\sqrt{s} = 13$ TeV with the ATLAS detector. Phys Rev D97(7):072016. https://doi.org/10.1103/PhysRevD.97.072016, arXiv:1712.08895
28. ATLAS Collaboration (2016) Search for the Standard Model Higgs boson decaying into $b\bar{b}$ produced in association with top quarks decaying hadronically in pp collisions at $\sqrt{s} = 8$ TeV with the ATLAS detector. JHEP 05:160. https://doi.org/10.1007/JHEP05(2016)160, arXiv:1604.03812
29. CMS Collaboration (2018) Search for $\mathrm{t\bar{t}H}$ production in the all-jet final state in proton-proton collisions at $\sqrt{s} = 13$ TeV. JHEP 06:101. https://doi.org/10.1007/JHEP06(2018)101, arXiv:1803.06986

Chapter 2
Theoretical Background

The search presented here and the main focus of my work has its roots in the standard model of particle physics. Particle physics is a branch of physics that attempts to describe the fundamental or elementary particles of the universe and their interactions. These elementary particles are the building blocks of everything humans have come to know, from physical objects, machinery and computing systems to biological organisms.

The notion that matter consists of indivisible particles was conceived around the 6th century BC by Greek philosophers. Of course the form and properties of these particles was not known until the 19th century, when atomic theory gained credence [1]. However the idea of the atom as the most fundamental particle was short lived, with the late 19th century discovery of the electron. Rutherford's discovery of the nucleus [2] in the early 20th century set the foundation of particle physics and paved the way to the discoveries of the proton and neutron, and many new particles in the 1950s. The idea that these new particles are formed from just a few elementary particles, called quarks, was independently proposed in 1964 by Gell-Mann [3] and Zweig [4].

The standard model of particle physics describes the known elementary particles and the fundamental forces governing their interactions. It was developed over the 1960s and 70s, and combines the theories of electromagnetic and weak interactions, and describes the strong interaction, i.e. quantum chromodynamics, as well as the Higgs mechanism.

In this chapter, descriptions of the theoretical models drawn upon in this work are provided. First, the structure of the standard model is introduced, followed by a detailed description of the physics behind the Higgs boson, and a discussion of its properties. Then, the measurements of the Higgs boson in high energy proton–proton collisions at the LHC are discussed, and finally the characteristics of the specific production and decay mode (channel) of this search are presented.

D. Salerno, *The Higgs Boson Produced With Top Quarks in Fully Hadronic Signatures*, Springer Theses, https://doi.org/10.1007/978-3-030-31257-2_2

2.1 The Standard Model of Particle Physics

The standard model of particle physics (SM) postulates the elementary particles that constitute matter, the fundamental forces of their interaction, the elementary particles which carry these forces, and the elementary particle that gives mass to the particles. The elementary particles of matter are known as fermions and are classified as either quarks or leptons. There are six types of quark, and six types of lepton, divided into three generations. The elementary particles that carry the fundamental forces are known as gauge bosons, of which there are four which carry the three fundamental forces described by the SM, namely the electromagnetic force, the weak force and the strong force. The final elementary particle in the SM is the Higgs boson, which is a scalar boson resulting from the mechanism which gives mass to the gauge bosons and fermions. The names and some properties (mass, charge and spin) of the 17 SM particles are given in Fig. 2.1.

The 12 elementary fermions of the SM can be divided into four distinct groups based on their electric charge Q, each with three particles from three generations:

- leptons with $Q = 0$: ν_e, ν_μ, ν_τ;
- leptons with $Q = -1$: e, μ, τ;
- quarks with $Q = 2/3$: u, c, t;
- quarks with $Q = -1/3$: d, s, b.

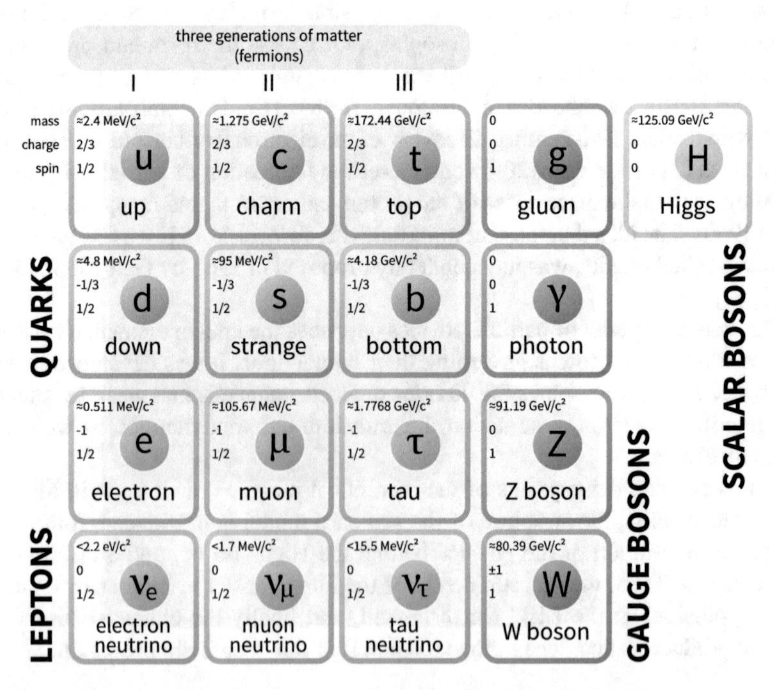

Fig. 2.1 Elementary particles of the SM: 12 fermions, 4 gauge bosons, and a scalar boson [5]. The shading indicates which forces interact with which fermion, as described below

The particles in each group are identical except for their mass, which defines the generation to which they belong, with higher masses in the higher generations. Each of these particles has an anti-particle which is equal in every way but for an opposite charge, however the anti-neutrinos are only distinguished by a more abstract quantity related to their spin. The six leptons exist freely and do not form bound states with other leptons in nature,[1] although the τ lepton is very short lived and decays into other fermions. On the other hand, the six quarks do not exist freely and must form bound states with other quarks to form one of two types of *hadrons*:

- mesons: one quark and one anti-quark, e.g. π^0, $\pi^\pm$, K^0, $K^\pm$, ρ, ϕ, J/ψ, ...;
- baryons: three quarks, e.g. p, n, Δ^0, $\Delta^\pm$, Λ^0, Λ_c^+, Λ_b^0, Σ^0, Ξ^0, ...

The mesons, being formed of two spin-$\frac{1}{2}$ particles, have integer spin and are thus classified as bosons, while the baryons have half-integer spin and are therefore fermions. The separation of a quark from its bound state requires a large amount of energy which is used to create new quark-antiquark pairs that form bound states with the original quarks in a process known as *fragmentation* and *hadronisation*. Most of these hadrons are unstable and decay, with a mean lifetime ranging from less than a microsecond to several minutes, to stable particles, namely electrons, protons and neutrinos.

The five bosons of the SM play the role of mediators of the fundamental interactions:

- the gluon g: massless, neutral, spin-1 gauge boson that mediates the strong interaction;
- the photon γ: massless, neutral, spin-1 gauge boson that mediates the electromagnetic interaction;
- the W boson $W^\pm$: massive, charged, spin-1 gauge bosons that mediate the charged-current weak interaction;
- the Z boson Z: massive, neutral, spin-1 gauge boson that mediates the neutral-current weak interaction;
- the Higgs boson H: massive, neutral spin-0 scalar boson that mediates interactions with the Higgs field (see Sect. 2.2).

Each of these interactions applies to some or all of the SM fermions, which means that some forces are only "felt" by some fermions, while other fermions are "immune" to them. The electromagnetic interaction applies to all charged fermions, i.e. it has no effect on neutrinos, and is responsible for many everyday phenomena such as electricity, magnetism, pressure and contact. The charged and neutral-current weak interactions apply to all 12 fermions and are responsible for nuclear decay and neutrino interactions with matter. At high energies, the electromagnetic and weak interactions unify to form the electroweak interaction. The strong interaction applies only to quarks. It is the force responsible for the confinement of quarks in bound states, and is so named since it is several orders of magnitude stronger than the electromagnetic and weak forces at short range.

[1] Laboratory bound states can and have been produced however, e.g. positronium, the unstable bound state of an electron and its anti-particle, the positron.

2.1.1 The SM Lagrangian

The SM is a quantum field theory which is invariant under local transformations of its gauge group:

$$G_{\mathrm{SM}} = \mathrm{SU}(3)_{\mathrm{C}} \otimes \mathrm{SU}(2)_{\mathrm{L}} \otimes \mathrm{U}(1)_{\mathrm{Y}}. \tag{2.1}$$

$\mathrm{SU}(3)_{\mathrm{C}}$, where the C stands for *colour*, is the symmetry group associated with the quantum chromodynamics (QCD) gauge theory [3, 4, 6–9], which describes the strong interaction between coloured quarks. $\mathrm{SU}(2)_{\mathrm{L}} \otimes \mathrm{U}(1)_{\mathrm{Y}}$ is the symmetry group of weak left-handed *isospin* and *hypercharge* associated to the electroweak theory developed by Glashow, Weinberg and Salam [10–12], which is a Yang-Mills theory [13] that describes the electromagnetic and weak interactions between quarks and leptons.

In a gauge theory, the number of generators is a characteristic of the symmetry group. In the SM, these generators are associated to the spin-1 gauge bosons previously described, except that there must be eight gauge bosons associated with the $\mathrm{SU}(3)_{\mathrm{C}}$ group, and therefore eight different colour states are assigned to the gluon. In total there are 12 generators for the three subgroups of Eq. (2.1), denoted $X_{1,2,\ldots,8}$, $T_{1,2,3}$, and Y, which are associated to the gauge fields G^a_μ with $a = 1, 2, \ldots, 8$, W^a_μ with $a = 1, 2, 3$, and B_μ, respectively. The Lagrangian density of the free gauge fields can be expressed in terms of the field strengths [14]:

$$G^a_{\mu\nu} = \partial_\mu G^a_\nu - \partial_\nu G^a_\mu + g_s f^{abc} G^b_\mu G^c_\nu \tag{2.2}$$

$$W^a_{\mu\nu} = \partial_\mu W^a_\nu - \partial_\nu W^a_\mu + g_2 \epsilon^{abc} W^b_\mu W^c_\nu \tag{2.3}$$

$$B_{\mu\nu} = \partial_\mu W_\nu - \partial_\nu W_\mu, \tag{2.4}$$

where f^{abc} and ϵ^{abc} are the respective structure constants of $\mathrm{SU}(3)_{\mathrm{C}}$ and $\mathrm{SU}(2)_{\mathrm{L}}$, and g_s, g_2, g_1 are adimensional coupling constants associated to the fields. g_1 is associated to the $\mathrm{U}(1)_{\mathrm{Y}}$ subgroup, which has no self coupling, and only appears in the interaction with matter fields.

As already mentioned, the SM describes the 12 elementary particles of matter, quarks and leptons. It does this through 45 matter fields, divided into three generations, each composed of 15 fields, known as Weyl spinors:

- an up and down left-handed doublet of quarks, $Q = (u_L, d_L)$, in 3 different colours;
- up and down right-handed singlets of quarks, u_R and d_R, in 3 different colours;
- an up and down left-handed doublet of leptons, $L = (\nu_L, e_L)$;
- a right-handed lepton singlet, e_R.

The spinors can be characterised in terms of their quantum numbers of the respective subgroups, as listed in Table 2.1. The first two columns indicate the transformation properties under the colour and isospin groups, respectively, while the last columns lists the hypercharge of each field. The hypercharge generator Y is related to the electric-charge generator Q (unfortunately the same notation as the left-handed quark doublet) by the relation $Q = T_3 + Y/2$.

Table 2.1 Gauge quantum numbers of a generation of SM fermions

	$SU(3)_C$	$SU(2)_L$	$U(1)_Y$
Q	3	2	1/3
u_R	3	1	4/3
d_R	3	1	−2/3
L	1	2	−1
e_R	1	1	−2

The matter fields, represented by the spinors ψ_L and ψ_R, are coupled to the gauge fields by the covariant derivate D_μ, which, for quarks that interact with all gauge fields, are defined as:

$$D_\mu \psi_L = \left(\partial_\mu - ig_s X_a G^a_\mu - ig_2 T_a W^a_\mu - ig_1 \frac{Y}{2} B_\mu\right)\psi_L, \tag{2.5}$$

$$D_\mu \psi_R = \left(\partial_\mu - ig_s X_a G^a_\mu - ig_1 \frac{Y}{2} B_\mu\right)\psi_R, \tag{2.6}$$

where $X_a = \frac{1}{2}\lambda_a$, $T_a = \frac{1}{2}\tau_a$, and Y are the generators of $SU(3)_C$, $SU(2)_L$ and $U(1)_Y$, respectively. λ_a are the Gell-Mann matrices [15] and τ_a are the Pauli matrices [16]. The commutation relations between the various generators are given by:

$$\begin{aligned} [X^a, X^b] &= if^{abc}X_c \quad \text{with} \quad \text{Tr}[X^a X^b] = \frac{1}{2}\delta_{ab}, \\ [T^a, T^b] &= i\epsilon^{abc}T_c, \\ [Y, Y] &= 0. \end{aligned} \tag{2.7}$$

The structure of the Lagrangian is determined by its invariance under local gauge transformations of $SU(3)_C \otimes SU(2)_L \otimes U(1)_Y$, which dictate the allowed combinations of the fields. For the electroweak sector, the local transformations act on the left-handed (L) and right-handed (R) fermions, and the gauge fields as follows:

$$\begin{aligned} L(x) &\to e^{i\alpha_a(x)T^a + i\beta(x)Y} L(x) \\ R(x) &\to e^{i\beta(x)Y} R(x) \\ \vec{W}_\mu(x) &\to \vec{W}_\mu(x) - \frac{1}{g_2}\partial_\mu \vec{\alpha}(x) - \vec{\alpha}(x) \times \vec{W}_\mu(x) \\ B_\mu(x) &\to B_\mu(x) - \frac{1}{g_1}\partial_\mu \beta(x), \end{aligned} \tag{2.8}$$

where $\alpha_a(x)$ and $\beta(x)$ are arbitrary functions. For the strong sector, which acts equally on left and right-handed fermions ψ, the local transformations take the form:

$$
\begin{aligned}
\psi(x) &\rightarrow e^{i\theta_a(x)X^a}\psi(x) \\
G_\mu^a(x) &\rightarrow G_\mu^a(x) - \frac{1}{g_s}\partial_\mu\theta^a(x) - f^{abc}\theta^b(x)G_\mu^c(x),
\end{aligned} \tag{2.9}
$$

where $\theta_a(x)$ is an arbitrary function.

Bringing it all together leads to a Lagrangian density which is invariant under local gauge transformations of Eq. (2.1), and consists of a free-field component and an interaction component. Ignoring mass terms for fermions and gauge bosons, the SM Lagrangian for a single generation can be written as:

$$
\begin{aligned}
\mathcal{L}_{\text{SM}} = &-\frac{1}{4}G_{\mu\nu}^a G_a^{\mu\nu} - \frac{1}{4}W_{\mu\nu}^a W_a^{\mu\nu} - \frac{1}{4}B_{\mu\nu}B^{\mu\nu} \\
&+ \bar{L}iD_\mu\gamma^\mu L + \bar{e}_R iD_\mu\gamma^\mu e_R \\
&+ \bar{Q}iD_\mu\gamma^\mu Q + \bar{u}_R iD_\mu\gamma^\mu u_R + \bar{d}_R iD_\mu\gamma^\mu d_R,
\end{aligned} \tag{2.10}
$$

where the covariant derivates D_μ contain only the relevant interactions from Eqs. (2.5) and (2.6). The first row of Eq. (2.10) describes the dynamics of the gauge fields and includes the kinetic (free) term, as well as triple and quartic self-interaction terms (not present for the abelian $\text{U}(1)_\text{Y}$ group), while the second and third rows contain the kinetic parts of the fermion fields plus their interactions with the gauge fields (where present).

Of course, the fermions and gauge bosons, with the exception of the neutrinos, gluons and photon, are experimentally proven to have mass, and therefore the SM Lagrangian must accommodate their mass terms. For the strong interaction, which is mediated by massless gluons, masses can be generated for the quarks while maintaining the gauge invariance under $\text{SU}(3)_\text{C}$, by adding terms of the form $-m\bar{\psi}\psi$. In the case of the electroweak interaction though, if mass terms for the gauge bosons of the form $\frac{1}{2}M^2W_\mu W^\mu$ are added to the Lagrangian, then the $\text{SU}(2)_\text{L} \otimes \text{U}(1)_\text{Y}$ gauge invariance would be violated. The problem of generating masses for the gauge bosons and fermions while maintaining the gauge invariance of $\text{SU}(2)_\text{L} \otimes \text{U}(1)_\text{Y}$ can be solved with the introduction of a mechanism of spontaneous symmetry breaking, described in the next section.

2.2 The Higgs Sector

The problem that known massive particles appear massless in the original gauge invariant Lagrangian was identified in the early 1960s. Several ideas emerged that aimed to solve this mass problem, which culminated in three independently developed models by Englert and Brout [17], Higgs [18], and Guralnik, Hagen and Kibble [19] in 1964.

In 1967, Weinberg [11] and Salam [12] independently demonstrated that the Higgs mechanism could be used to break the symmetry of the electroweak model developed by Glashow [10]. Weinberg additionally observed that this would also provide mass terms for the fermions. These theories were somewhat neglected by the scientific community of the time, and it was not until 't Hooft published his work on renormalisable models [20] in 1971, that the Higgs mechanism gained widespread acceptance. With the discovery of the top quark in 1995 [21, 22], all particles predicted by the SM, except the Higgs boson, had been observed. This missing piece related to the Higgs sector then became the topic of central importance to particle physics and dominated the search program at LEP during its final stages. The breakthrough finally arrived in 2012 when the ATLAS and CMS experiments at the LHC confirmed the discovery of a new boson consistent with the SM Higgs boson [23, 24].

In this section, the fundamental theories underlying the mechanism of spontaneous symmetry breaking are described. Some details of electroweak interactions are briefly discussed and then the Higgs boson in the context of the SM is explained.

2.2.1 *The Brout–Englert–Higgs Mechanism*

The Brout–Englert–Higgs–Guralnik–Hagen–Kibble mechanism of spontaneous symmetry breaking [17–19], hereafter referred to as the Higgs mechanism, built upon the ideas developed in the Goldstone model [25]. It was then further built upon by Weinberg and Salam, who combined it with Glashow's model and applied it to the SM. All three methods are outlined in the following.

The Goldstone model

Consider a complex scalar field $\phi(x) = \dfrac{\phi_1(x) + i\phi_2(x)}{\sqrt{2}}$ with a quartic interaction. The Lagrangian density is given by:

$$\mathcal{L} = \partial^\mu \phi^\dagger \partial_\mu \phi - V(\phi) \,, \qquad V(\phi) = \mu^2 \phi^\dagger \phi + \lambda (\phi^\dagger \phi)^2. \tag{2.11}$$

This Lagrangian is invariant under global U(1) phase transformations of the field:

$$\phi \to \phi' = e^{i\alpha}\phi \,, \qquad \phi^\dagger \to \phi^{\dagger\prime} = e^{-i\alpha}\phi^\dagger, \tag{2.12}$$

and the corresponding Hamiltonian density is:

$$\mathcal{H} = \partial^0 \phi^\dagger \partial_0 \phi + (\nabla \phi^\dagger) \cdot (\nabla \phi) + V(\phi). \tag{2.13}$$

The stability of the theory requires the potential energy to be bounded from below, and therefore $\lambda > 0$ in Eq. (2.11). On quantisation, the configuration that minimises the energy corresponds to the vacuum state $\langle 0|\phi|0\rangle$. Since the first two terms of

Eq. (2.13) are positive definite and vanish for constant $\phi(x)$, the minimum value of $\mathcal{H}$ corresponds to the constant $\phi(x)$ that minimises $V(\phi)$. This leads to two different situations depending on the sign of μ^2:

1. $\mu^2 > 0$: In this case, $V(\phi)$ has an absolute minimum at $\phi(x) = 0$ as shown in Fig. 2.2a. Ignoring the quartic term $\lambda|\phi|^4$, Eq. (2.11) reduces to the Lagrangian of a complex Klein–Gordon field. On quantisation, this gives rise to charged spin-0 particles of mass μ with a unique vacuum state of $\langle 0|\phi(x)|0\rangle = 0$ that is symmetric under the transformations (2.12). The quartic term can be treated with perturbation theory and represents a self interaction of the particle.
2. $\mu^2 < 0$: The form of the potential in this case is shown in Fig. 2.2b. While $\phi = 0$ corresponds to a local maximum, the minimum of $V(\phi)$ occurs at all the points along the circle:

$$\phi(x) = \phi_0 = \sqrt{\frac{-\mu^2}{2\lambda}}e^{i\theta}, \tag{2.14}$$

where θ is the phase angle in the complex plane. The vacuum state in this case is not unique and no point within it is symmetric under the transformations (2.12). However, introducing a driving term to the potential of the form $-\epsilon\phi^\dagger - \epsilon^*\phi^\dagger$ forces $V(\phi)$ to have a unique minimum with the same (arbitrary) phase as ϵ. Choosing the phase to be $\theta = 0$ and taking $\epsilon \to 0$, the minimum moves on to the circle (2.14) and takes the real value:

$$\phi_0 = \sqrt{\frac{-\mu^2}{2\lambda}} = \frac{1}{\sqrt{2}}v \quad (> 0), \tag{2.15}$$

where $v = \sqrt{-\mu^2/\lambda}$. Upon quantisation, the configuration of minimum energy corresponds to a non-zero vacuum expectation value of:

$$\langle 0|\phi(x)|0\rangle = \phi_0 = \frac{1}{\sqrt{2}}v \neq 0, \tag{2.16}$$

which is not symmetric. This non-zero, non-symmetric vacuum expectation value is the condition for spontaneous symmetry breaking.

The complex field can be expanded around the ground state configuration by introducing two real fields $\sigma(x)$ and $\eta(x)$ which represent the deviations from its equilibrium:

$$\phi(x) = \frac{v + \sigma(x) + i\eta(x)}{\sqrt{2}}. \tag{2.17}$$

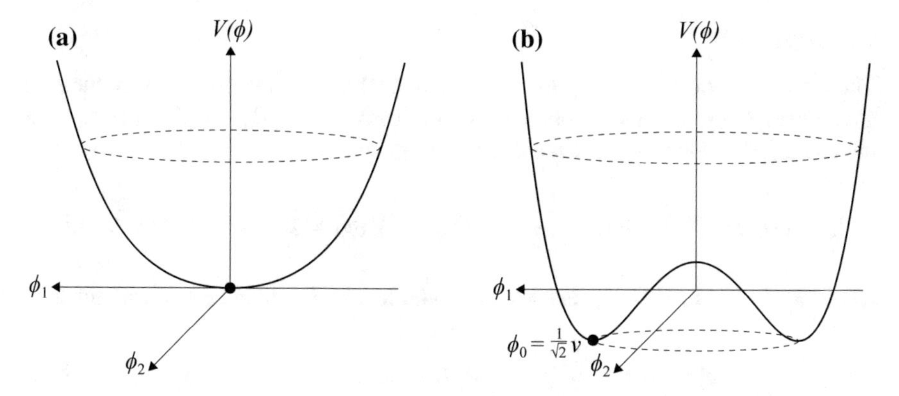

Fig. 2.2 The potential energy density $V(\phi) = \mu^2|\phi|^2 + \lambda|\phi|^4$ for $\lambda > 0$ and $\mu^2 > 0$ (**a**) and $\mu^2 < 0$ (**b**)

Substituting this expression for ϕ into the Lagrangian density (2.11) yields:

$$\mathcal{L} = \frac{1}{2}\partial^\mu\sigma\partial_\mu\sigma - \frac{1}{2}(2\lambda v^2)\sigma^2 + \frac{1}{2}\partial^\mu\eta\partial_\mu\eta \\ - \lambda v\sigma(\sigma^2+\eta^2) - \frac{1}{4}\lambda(\sigma^2+\eta^2)^2 + \text{const.} \tag{2.18}$$

The quadratic terms in σ and η lead to the free Lagrangian density:

$$\mathcal{L}_0 = \frac{1}{2}\partial^\mu\sigma\partial_\mu\sigma - \frac{1}{2}(2\lambda v^2)\sigma^2 + \frac{1}{2}\partial^\mu\eta\partial_\mu\eta, \tag{2.19}$$

while the constant is insignificant and the remaining terms (those cubic and quartic in σ and η) represent interactions. Equation (2.19) describes two real Klein–Gordon fields, which upon quantisation lead to neutral spin-0 particles: the σ boson of mass $\sqrt{2\lambda v^2}$ and the η boson of zero mass (since there is no term in η^2). This mass spectrum is a consequence of the spontaneous symmetry breaking. Consider small displacements from the equilibrium configuration $\phi(x) = \phi_0$ (see Fig. 2.2b) where $\sigma(x)$ represents a displacement in the radial plane $\phi_2 = 0$ in which $V(\phi)$ increases quadratically, while $\eta(x)$ represents a displacement along the circle of minimum potential in which $V(\phi)$ is constant. The quantum excitation of the $\eta(x)$ field—the η boson—is therefore massless.

The presence of a zero-mass particle is a consequence of the degeneracy of the vacuum. The fact that spontaneous symmetry breaking leads to massless bosons is the essence of the *Goldstone Theorem* and the zero-mass particles are known as *Goldstone bosons* [26].

The Higgs model

The Goldstone model can be generalised to a local U(1) symmetry by introducing a gauge field $A_\mu(x)$ and using the covariant derivative $D_\mu = \partial_\mu - ieA_\mu$, where $-e$ is the charge of an electron. The resultant Lagrangian,

$$\mathcal{L}_0 = (D_\mu\phi)^\dagger D^\mu\phi - V(\phi) - \frac{1}{4}F_{\mu\nu}F^{\mu\nu}, \qquad V(\phi) = \mu^2\phi^\dagger\phi + \lambda(\phi^\dagger\phi)^2, \quad (2.20)$$

where $F_{\mu\nu} = \partial_\nu A_\mu - \partial_\mu A_\nu$, is invariant under local U(1) gauge transformations:

$$\phi(x) \to e^{i\alpha(x)}\phi(x), \qquad A_\mu(x) \to A_\mu(x) + \frac{1}{e}\partial_\mu\alpha(x). \quad (2.21)$$

As in the Goldstone model, the stability of the theory requires $\lambda > 0$ and two situations arise depending on the sign of μ^2.

1. For $\mu^2 > 0$, the state of minimum energy corresponds to $\phi = 0$ and $A_\mu = 0$ and, on quantisation, Eq. (2.20) describes a charged scalar particle (and its anti-particle) of mass μ and a massless spin-1 boson (with two polarisation states). The total number of degrees of freedom in this case is four: two for each field.
2. For $\mu^2 < 0$, the vacuum state is not unique and spontaneous symmetry breaking occurs. The vector field A_μ vanishes in the vacuum but the scalar field takes on a non-zero value given by (2.14). As for the Goldstone model, a unique minimum can be imposed with a driving term resulting in the real value (2.15) for ϕ_0. As previously, the field ϕ can be expanded in terms of the real fields $\sigma(x)$ and $\eta(x)$ defined by Eq. (2.17). The Lagrangian density (2.20), after omitting some higher order interaction terms and an insignificant constant, then becomes:

$$\begin{aligned}\mathcal{L}_0 = &\frac{1}{2}\partial_\mu\sigma\partial^\mu\sigma - \frac{1}{2}(2\lambda v^2)\sigma^2 \\ &- \frac{1}{4}F_{\mu\nu}F^{\mu\nu} + \frac{1}{2}(ev)^2A_\mu A^\mu + \frac{1}{2}\partial_\mu\eta\partial^\mu\eta - evA^\mu\partial_\mu\eta. \end{aligned} \quad (2.22)$$

A direct interpretation of the Lagrangian highlights some inconsistencies. The first line of Eq. (2.22) describes a real Klein–Gordon field which upon quantisation corresponds to an uncharged spin-0 boson of mass $\sqrt{2\lambda v^2}$. The second line however, includes the term $A^\mu\partial_\mu\eta$ which prevents the interpretation of a massive vector boson with mass ev and a massless scalar boson. This inconsistency also arises considering the number of degrees of freedom. In the case of $\mu^2 > 0$, Eq. (2.20) has four degrees of freedom: two from the complex scalar field $\phi(x)$ and two from the real massless vector field $A_\mu(x)$. On the other hand, Eq. (2.22) apparently gives five degrees of freedom: one for each of the real scalar fields $\sigma(x)$ and $\eta(x)$, and three for the massive vector field $A_\mu(x)$. Since the number of degrees of freedom must be conserved, one of the fields must be non-physical and can therefore be eliminated.

In fact, the scalar field $\eta(x)$ can be eliminated all together by a suitable choice of gauge transformation of the form (2.21). The gauge which transforms the complex field $\phi(x)$ into a real field of the form:

$$\phi(x) = \frac{v + \sigma(x)}{\sqrt{2}}, \tag{2.23}$$

is called the *unitary gauge*. Substituting (2.23) into the Lagrangian (2.20) leads to a Lagrangian of the form $\mathcal{L} = \mathcal{L}_0 + \mathcal{L}_\mathrm{I}$, where

$$\mathcal{L}_0 = \frac{1}{2}\partial_\mu \sigma \partial^\mu \sigma - \frac{1}{2}(2\lambda v^2)\sigma^2 - \frac{1}{4}F_{\mu\nu}F^{\mu\nu} + \frac{1}{2}(ev)^2 A_\mu A^\mu \tag{2.24}$$

is the free part containing the quadratic terms, and $\mathcal{L}_\mathrm{I}$ is the interaction component containing higher order interaction terms. As $\mathcal{L}_0$ contains no terms that couple $\sigma(x)$ and $A_\mu(x)$, they can be interpreted as a real Klein–Gordon field and a real massive vector field, respectively. On quantisation, $\sigma(x)$ gives rise to a neutral scalar boson of mass $M_\sigma = \sqrt{2\lambda v^2} = \sqrt{-2\mu^2}$, and $A_\mu(x)$ to a neutral vector boson of mass $M_A = ev = -e\mu^2/\lambda$. The total number of degrees of freedom is now four. One of the two degrees of freedom of the complex field $\phi(x)$ goes to the real field $\sigma(x)$ while the other is taken by the vector field $A_\mu(x)$, which has acquired mass.

The process by which a vector boson acquires mass without destroying the gauge invariance of the Lagrangian density is known as the *Higgs mechanism*, and the massive spin-0 boson associated with the field $\sigma(x)$ is called the *Higgs boson*. The field associated with the would-be Goldstone boson, $\eta(x)$, is eliminated by gauge invariance and its degree of freedom is transferred to the vector field $A_\mu(x)$.

The Glashow–Weinberg–Salam theory

In order to generate masses for the $W^\pm$ and Z bosons whilst maintaining a massless photon, the SU(2) symmetry must be broken and the U(1) symmetry must remain exact:

$$\mathrm{SU(2)_L} \otimes \mathrm{U(1)_Y} \rightarrow \mathrm{U(1)_Q}. \tag{2.25}$$

The choice of Weinberg and Salam was to introduce an SU(2) isospin doublet of scalar fields with weak hypercharge $Y = +1$:

$$\Phi = \begin{pmatrix} \phi^+ \\ \phi^0 \end{pmatrix}. \tag{2.26}$$

The Lagrangian can be written as the sum of the electroweak component of Eq. (2.10),

$$\mathcal{L}_\mathrm{ew} = -\frac{1}{4}W^a_{\mu\nu}W^{\mu\nu}_a - \frac{1}{4}B_{\mu\nu}B^{\mu\nu} + \bar{L}iD_\mu\gamma^\mu L + \bar{e}_R i D_\mu \gamma^\mu e_R + \dots, \tag{2.27}$$

and the SU(2) $\otimes$ U(1) gauge-invariant form of the scalar field component,

$$\mathcal{L}_H = (D_\mu \Phi)^\dagger (D^\mu \Phi) - \mu^2 \Phi^\dagger \Phi - \lambda (\Phi^\dagger \Phi)^2. \tag{2.28}$$

Here the covariant derivative for left-handed spinors and the scalar doublet is defined as $D_\mu = \partial_\mu - ig_2 \frac{1}{2}\tau_a W^a_\mu - ig_1 \frac{1}{2} B_\mu$, where g_1 and g_2 are coupling constants and $\tau_{1,2,3}$ are the Pauli matrices. For right-handed spinors, D_μ is defined without the term in W^a_μ.

As in the Higgs model, Φ assumes a vacuum expectation value:

$$\Phi_0 = \langle 0|\Phi|0\rangle = \begin{pmatrix} 0 \\ v/\sqrt{2} \end{pmatrix}, \qquad v = \sqrt{\frac{-\mu^2}{\lambda}}, \tag{2.29}$$

which preserves the $U(1)_Q$ gauge invariance:

$$e^{i\alpha Q}\Phi_0 = \begin{pmatrix} e^{i\alpha} & 0 \\ 0 & 1 \end{pmatrix} \begin{pmatrix} 0 \\ v/\sqrt{2} \end{pmatrix} = \Phi_0, \tag{2.30}$$

where the matrix $Q = \begin{pmatrix} +1 & 0 \\ 0 & 0 \end{pmatrix}$ represents the electric-charge generator acting on Φ.

The field $\Phi(x)$ can be expanded in terms of its deviations from the vacuum field Φ_0 using four real fields $\eta_{1,2,3}(x)$ and $\sigma(x)$:

$$\Phi(x) = \frac{1}{\sqrt{2}} \begin{pmatrix} \eta_1(x) + i\eta_2(x) \\ v + \sigma(x) + i\eta_3(x) \end{pmatrix}. \tag{2.31}$$

As in the Higgs model, three of these four fields are found to be non-physical and can be removed all together by employing the unitary gauge. The resultant field,

$$\Phi(x) = \frac{1}{\sqrt{2}} \begin{pmatrix} 0 \\ v + \sigma(x) \end{pmatrix}, \tag{2.32}$$

is expressed in terms of just one physical field $\sigma(x)$ corresponding to the neutral scalar Higgs boson. Substituting this expression for Φ into the Lagrangian (2.28) yields:

$$\begin{aligned} \mathcal{L}_H =& \frac{1}{2}\partial_\mu \sigma \partial^\mu \sigma - \frac{1}{2}\mu^2 (v+\sigma)^2 - \frac{1}{4}\lambda (v+\sigma)^4 \\ &+ \frac{1}{8} g_2^2 (v+\sigma)^2 |W^1_\mu - iW^2_\mu|^2 + \frac{1}{8}(v+\sigma)^2 |g_2 W^3_\mu - g_1 B_\mu|^2. \end{aligned} \tag{2.33}$$

Defining the physical fields $W^\pm_\mu$, Z_μ and A_μ as:

$$W^\pm_\mu = \frac{1}{\sqrt{2}}(W^1_\mu \mp iW^2_\mu), \qquad Z_\mu = \frac{g_2 W^3_\mu - g_1 B_\mu}{\sqrt{g_2^2 + g_1^2}}, \qquad A_\mu = \frac{g_2 W^3_\mu + g_1 B_\mu}{\sqrt{g_2^2 + g_1^2}}, \tag{2.34}$$

allows a straightforward identification of the Lagrangian (2.33), where the quadratic terms in the fields describe the particle masses:

$$\frac{1}{2}M_W^2 W_\mu^+ W^{-\mu} + \frac{1}{2}M_Z^2 Z_\mu Z^\mu + \frac{1}{2}M_A^2 A_\mu A^\mu. \tag{2.35}$$

Making this identification and comparing Eqs. (2.35) to (2.33) gives rise to massive W and Z bosons while maintaining massless photons:

$$M_W = \frac{1}{2}g_2 v, \qquad M_Z = \frac{1}{2}v\sqrt{g_2^2 + g_1^2}, \qquad M_A = 0. \tag{2.36}$$

The spontaneous symmetry breaking $\mathrm{SU(2)_L} \otimes \mathrm{U(1)_Y} \rightarrow \mathrm{U(1)_Q}$ has led to the $\mathrm{W}^\pm$ and Z bosons acquiring mass, while the $\mathrm{U(1)_Q}$ symmetry is still unbroken and so its generator, the photon, remains massless. The massive, electrically neutral, spin-0 Higgs boson corresponding to the field $\sigma(x)$ also remains.

In order to generate masses for the fermions, the same scalar field Φ and the isodoublet $\tilde{\Phi} = i\tau_2\Phi^*$ are coupled to the fermion fields in an $\mathrm{SU(2)_L} \otimes \mathrm{U(1)_Y}$ invariant Yukawa Lagrangian of the form:

$$\mathcal{L}_F = -g_e \bar{L}\Phi e_R - g_d \bar{Q}\Phi d_R - g_u \bar{Q}\tilde{\Phi} u_R + \text{h.c.}\,. \tag{2.37}$$

In general, terms of the form $g\bar{\psi}\phi\psi$, where ψ is a spinor and ϕ is a scalar field, are known as Yukawa interactions, with the constant g known as a Yukawa coupling. Substituting the expanded form (2.32) of the field Φ into Eq. (2.37) yields:

$$\begin{aligned}\mathcal{L}_F =& -\frac{1}{\sqrt{2}}g_e(\bar{\nu}_e, \bar{e}_L)\begin{pmatrix}0\\ v+\sigma\end{pmatrix}e_R - \frac{1}{\sqrt{2}}g_d(\bar{u}_L, \bar{d}_L)\begin{pmatrix}0\\ v+\sigma\end{pmatrix}d_R \\ & -\frac{1}{\sqrt{2}}g_u(\bar{u}_L, \bar{d}_L)\begin{pmatrix}v+\sigma\\ 0\end{pmatrix}u_R + \text{h.c.} \\ =& -\frac{1}{\sqrt{2}}g_e(v+\sigma)\bar{e}e - \frac{1}{\sqrt{2}}g_d(v+\sigma)\bar{d}d - \frac{1}{\sqrt{2}}g_u(v+\sigma)\bar{u}u,\end{aligned} \tag{2.38}$$

where $\bar{e}e = \bar{e}_L e_R + \bar{e}_R e_L$, $\bar{d}d = \bar{d}_L d_R + \bar{d}_R d_L$ and $\bar{u}u = \bar{u}_L u_R + \bar{u}_R u_L$. Constant terms in the Lagrangian (2.38) in front of $\bar{e}e$, $\bar{d}d$, and $\bar{u}u$ can be identified with the fermion masses, which leads to the following relations between Yukawa couplings and masses:

$$m_e = \frac{g_e v}{\sqrt{2}}, \qquad m_d = \frac{g_d v}{\sqrt{2}}, \qquad m_u = \frac{g_u v}{\sqrt{2}}. \tag{2.39}$$

The Yukawa coupling of a fermion is therefore proportional to its mass and given by:

$$g_f = \sqrt{2}\frac{m_f}{v}. \tag{2.40}$$

In summary, the Higgs isodoublet Φ of scalar fields is able to generate the masses of both the weak vector bosons, $W^{\pm}$ and Z, as well as the fermions, by spontaneously breaking the $SU(2)_L \otimes U(1)_Y$ gauge symmetry. On the other hand, the electromagnetic $U(1)_Q$ symmetry and the strong $SU(3)_C$ colour symmetry remain unbroken. The SM is said to exhibit $SU(3)_C \otimes SU(2)_L \otimes U(1)_Y$ gauge invariance when combined with the electroweak symmetry breaking mechanism.

2.2.2 Electroweak Interactions

The rotations that diagonalise the mass matrix of the gauge bosons, leading to the physical fields defined in Eq. (2.34), also define the electroweak mixing angle:

$$\sin\theta_W = \frac{g_1}{\sqrt{g_1^2 + g_2^2}} = \frac{e}{g_2}, \tag{2.41}$$

which can be expressed in terms of the W and Z boson masses as:

$$\sin^2\theta_W = 1 - \cos^2\theta_W = 1 - \frac{M_W^2}{M_Z^2}. \tag{2.42}$$

Expanding the covariant derivative in $\mathcal{L}_{SM}$, given in Eq. (2.10), and rewriting the components containing fermion fields in terms of the new fields $W_\mu^{\pm}$, Z_μ and A_μ, leads to the neutral and charged-current Lagrangians:

$$\begin{aligned} \mathcal{L}_{NC} &= eJ_\mu^A A^\mu + \frac{g_2}{\cos\theta_w} J_\mu^Z Z^\mu, \\ \mathcal{L}_{CC} &= \frac{g_2}{\sqrt{2}}(J_\mu^+ W^{+\mu} - J_\mu^- W^{-\mu}), \end{aligned} \tag{2.43}$$

where the currents J_μ are given by:

$$\begin{aligned} J_\mu^A &= Q_f \bar{f}\gamma_\mu f, \\ J_\mu^Z &= \frac{1}{2}\bar{f}\gamma_\mu\left[(T_f^3 - 2Q_f\sin^2\theta_W) - \gamma_5(T_f^3)\right] f, \\ J_\mu^+ &= \frac{1}{2}\bar{f}_u\gamma_\mu(1-\gamma_5) f_d. \end{aligned} \tag{2.44}$$

Here f_u and f_d are the up and down-type fermions, while f refers to either of them, and Q_f is the electric charge of the fermion.

The results presented so far are only valid for one generation of leptons and quarks. When all three generations of leptons and quarks are considered, the couplings g_e, g_d, and g_u of Eq. (2.37) become unitary matrices in the "generation space". By applying an SU(3) rotation, g_e can be made diagonal, however this can only happen for one of g_u and g_d, but not both, implying that the mass eigenstates for the quarks q are not identical to the current eigenstates q'. Allowing u-type quarks to have equal mass and current eigenstates, means that the d-type quark current eigenstates are a mixture of d-type quark mass eigenstates, connected by a unitary matrix, known as the Cabibbo–Kobayashi–Maskawa (CKM) matrix [27, 28], V_{CKM}:

$$q' = V_{\text{CKM}} q \quad \Rightarrow \quad \begin{bmatrix} d' \\ s' \\ b' \end{bmatrix} = \begin{bmatrix} V_{ud} & V_{us} & V_{ub} \\ V_{cd} & V_{cs} & V_{cb} \\ V_{td} & V_{ts} & V_{tb} \end{bmatrix} \begin{bmatrix} d \\ s \\ b \end{bmatrix}. \tag{2.45}$$

The square of the individual elements of the matrix $|V_{ij}|^2$ is proportional to the probability of a transition from quark i to quark j. The unitarity of V_{CKM} ensures that the neutral currents are diagonal for both types of quark, and thus mixing of quark flavours only occurs for the charged-current weak interaction. The absence of flavour-changing neutral currents in the SM was first explained by Glashow, Iliopoulos, and Maiani [29] via what is now known as the GIM mechanism. In the case of leptons, the SM assumption of massless neutrinos ensures that the mass and current eigenstates coincide, and no such mixing occurs.

2.2.3 *The Standard Model Higgs Boson*

Although the Higgs boson plays an important role in the SM, its mass is not set by the theory and remains a free parameter. In fact, extracting the terms containing the Higgs field $\sigma(x)$ from the Lagrangian density (2.33) leads to:

$$\mathcal{L}_H = \frac{1}{2}\partial_\mu \sigma \partial^\mu \sigma - \lambda v^2 \sigma^2 - \lambda v \sigma^3 - \frac{1}{4}\lambda \sigma^4. \tag{2.46}$$

This equation simply identifies the Higgs boson mass as:

$$M_H^2 = 2\lambda v^2 = -2\mu^2. \tag{2.47}$$

Equation (2.46) also provides the Higgs self-interaction couplings by using the Feynman rules[2]:

$$g_{H^3} = (3!) i \lambda v = 3i \frac{M_H^2}{v}, \qquad g_{H^4} = (4!) i \frac{1}{4} \lambda = 3i \frac{M_H^2}{v^2}. \tag{2.48}$$

Similarly, the Higgs couplings to gauge bosons (V) and fermions (f) are provided by the relevant terms in the Lagrangian (2.33) and (2.38), respectively:

---

[2]The Feynman rules applied to these couplings: multiply the term involving the interaction by a factor $-i$ and $n!$, where n is the number of identical particles in the vertex.

$$\mathcal{L}_{M_V} \sim M_V^2 \left(1+\frac{\sigma}{v}\right)^2 V^\dagger V, \qquad \mathcal{L}_{m_f} \sim -m_f \left(1+\frac{\sigma}{v}\right) \bar{f} f, \tag{2.49}$$

which lead to the following expressions for the couplings:

$$g_{HVV} = -2\frac{M_V^2}{v}, \qquad g_{HHVV} = -2\frac{M_V^2}{v^2}, \qquad g_{Hff} = \frac{m_f}{v}. \tag{2.50}$$

Equation (2.50) implies that the Higgs boson couples to fermions with strength proportional to their mass, and to gauge bosons with strength proportional to their mass squared.

The vacuum expectation value of the Higgs field v can be linked to the Fermi coupling constant G_F via the W boson mass[3] to obtain a numerical value:

$$M_W = \frac{1}{2} g_2 v = \left(\frac{\sqrt{2} g_2^2}{8 G_F}\right)^{1/2} \quad \Rightarrow \quad v = \frac{1}{(\sqrt{2} G_F)^{1/2}} \simeq 246\,\text{GeV}. \tag{2.51}$$

Since the Higgs coupling to bosons and fermions does not depend on M_H (due to the unknown value of λ), and neither does the vacuum expectation value, the Higgs mass is a free parameter of the SM. However, theoretical limits can be placed on M_H and, since its discovery, increasingly precise measurements of its mass have been made. Moreover, the SM predicts several properties of the Higgs boson which have also been measured with increasing accuracy in recent years. The theoretical mass limits and predicted properties of the Higgs boson, as well as recent experimental measurements, are discussed in Sect. 2.3.

2.3 Higgs Boson Properties

2.3.1 Theoretical Considerations

Not only is the Higgs boson mass a free parameter of the SM, but so are the fermion masses, since the number of Yukawa couplings introduced in Eq. (2.39) is equal to the number of masses. On the other hand, the fermion couplings to gauge bosons are predicted by the theory, as non-trivial functions of the fermion quantum numbers of left-handed weak isospin T_f^3 and electric charge Q_f, and the weak mixing angle $\sin\theta_W$. From Eqs. (2.43) and (2.44), the vector and axial-vector couplings of the fermion f to the Z boson can be written as:

$$v_f = \frac{T_f^3 - 2Q_f \sin^2\theta_W}{2\sin\theta_W \cos\theta_W}, \qquad a_f = \frac{T_f^3}{2\sin\theta_W \cos\theta_W}, \tag{2.52}$$

[3] $G_F = 1.166 \times 10^{-5}\,\text{GeV}^{-2}$ is experimentally determined from muon decays mediated by the W boson.

while the couplings to the W boson are simply:

$$v_f = a_f = \frac{1}{2\sqrt{2}\sin\theta_W}. \tag{2.53}$$

In a similar fashion, the trilinear coupling between the electroweak gauge bosons can be derived from the appropriate Lagrangians, and are given by:

$$g_{WWA} = g_2 \sin\theta_W = e, \qquad g_{WWZ} = e\cos\theta_W / \sin\theta_W, \tag{2.54}$$

where e is the electric charge.

The relative strength of the neutral and charged currents, $J_Z^{\mu} J_{\mu Z} / J^{\mu +} J_{\mu}^{-}$, can be determined with the parameter ρ, which is given by:

$$\rho = \frac{M_W^2}{\cos^2\theta_W M_Z^2}, \tag{2.55}$$

and is equal to one in the SM, by Eq. (2.42). This is a consequence of the doublet nature of the Higgs field, and would not hold if the Higgs multiplet was composed of three or more fields.

The SM Higgs boson is a CP-even spin-0 scalar which is its own antiparticle and therefore is assigned the quantum numbers $J^{PC} = 0^{++}$. As already discussed in Sect. 2.2, the SM Higgs boson has no charge and does not experience the strong interaction, i.e. it has no colour charge. Its mass is a free parameter of the SM, but is related to its self coupling λ and the vacuum expectation value v via the relation $m_H = \sqrt{2\lambda}v$. Furthermore, the measured Higgs boson mass of approximately 125 GeV fits nicely with some important theoretical considerations, as described in the following.

Due to quantum corrections, the coupling constants and the masses appearing in the SM depend on the considered energy scale, Q^2, which leads to the so-called running coupling constants. This is also the case for the quartic Higgs coupling which is monotonically increasing with the energy scale: $\lambda(Q^2)$.

Consider the one-loop radiative corrections to the Higgs boson quartic self-coupling as shown in Fig. 2.3. The running constant λ can be written in terms of the energy scale Q and the electroweak symmetry breaking scale (or vacuum expectation value) v as:

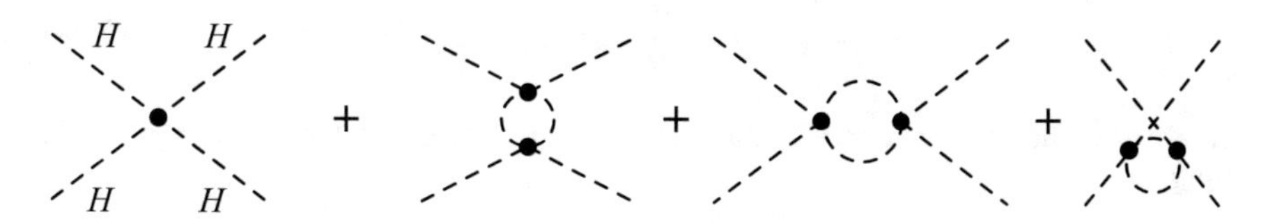

Fig. 2.3 Typical Feynman diagrams for the tree-level and one-loop Higgs self-coupling

$$\lambda(Q^2) = \lambda(v^2)\left[1 - \frac{3}{4\pi^2}\lambda(v^2)\log\frac{Q^2}{v^2}\right]^{-1}. \tag{2.56}$$

If the energy is much smaller than the electroweak breaking scale, $Q^2 \ll v^2$, the quartic coupling becomes extremely small and eventually vanishes, $\lambda(Q^2) \sim \lambda(v^2)/(1 - \log(0)) \rightarrow 0_+$. The theory is said to be trivial, i.e. non-interacting since the coupling is zero.

On the other hand, if the energy is much higher than the weak scale, $Q^2 \gg v^2$, the quartic coupling grows and at a certain point becomes infinite, $\lambda(Q^2) \sim \lambda(v^2)/(1 - \log(e)) \rightarrow \infty$. The point where the coupling becomes infinite, is called the Landau pole, and occurs when:

$$Q \rightarrow \Lambda_C = v\exp\left(\frac{2\pi^2}{3\lambda}\right) = v\exp\left(\frac{4\pi^2 v^2}{3M_{\mathrm{H}}^2}\right). \tag{2.57}$$

Below this cut-off energy scale Λ_C, the self-coupling λ remains finite and the SM retains its validity. Equation (2.57) provides an approximate constraint on the cut-off energy scale, for example with $M_{\mathrm{H}} \approx 125$ GeV then $\Lambda_C \sim 10^{24}$ GeV.

In addition to the Higgs boson loops, the running coupling constant is also affected by contributions from fermions and gauge bosons. Since the Higgs boson couplings are proportional to the particle masses, the contributions from the top quark and the massive gauge bosons are dominant. With the measured values of the masses of the Higgs boson, top quark, W and Z bosons, the Higgs quartic coupling remains perturbative all the way up to the Planck scale. In fact, given all available measurements of SM parameters, the gauge couplings and the Yukawa couplings also remain perturbative all the way up to $M_{\mathrm{Planck}} \approx 2.4 \times 10^{18}$ GeV, thus rendering the SM a consistent, calculable theory.

2.3.2 Higgs Boson Decays

At leading order (LO), the Higgs boson decay to fermions, $\mathrm{H}(p_h) \rightarrow f(p_1)\bar{f}(p_2)$, is represented by the Feynman diagram in Fig. 2.4a, and the transition amplitude of the process is given by:

$$\mathcal{M}_{H\rightarrow f\bar{f}} = \frac{-i}{v} m_f \bar{u}(p_1)v(p_2), \tag{2.58}$$

where v is the vacuum expectation value. The spin-averaged amplitude squared is then:

$$\sum_{\mathrm{spin}} |\mathcal{M}_{H\rightarrow f\bar{f}}|^2 = \frac{m_f^2}{v^2}\mathrm{Tr}\left[(\not{p}_1 + m_f)(\not{p}_2 - m_f)\right] = \frac{4m_f^2}{v^2}(p_1 p_2 - m_f^2). \tag{2.59}$$

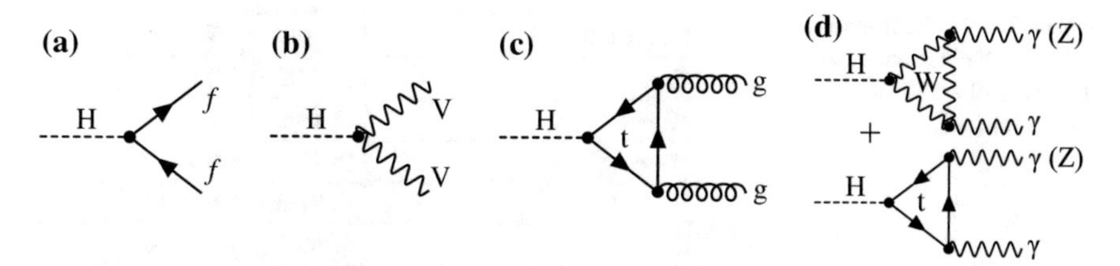

Fig. 2.4 Feynman diagrams of the LO Higgs boson decay processes: (**a**) Decays to fermions. (**b**) Decays to weak vector bosons (V = W, Z). (**c**) Decays to gluons. (**d**) Decays to photons or Zγ

In the rest frame of the Higgs boson, the four momenta are given by:

$$p_h^\mu = (M_H, \vec{0}), \qquad p_1^\mu = (E_f, \vec{p}), \qquad p_2^\mu = (E_f, -\vec{p}), \tag{2.60}$$

and conservation of energy requires $M_H = 2E_f$. Setting $p = |\vec{p}|$ implies $E_f^2 = p^2 + m_f^2$, which leads to the following relations:

$$p_1 p_2 - m_f^2 = \frac{1}{2}M_H^2\left(1 - \frac{4m_f^2}{M_H^2}\right), \qquad p = \frac{1}{2}M_H\sqrt{1 - \frac{4m_f^2}{m_H^2}}. \tag{2.61}$$

The amplitude squared can then be written as:

$$\sum_{\text{spin}} |\mathcal{M}_{H\to f\bar{f}}|^2 = N_C \frac{2m_f^2}{v^2} M_H^2 \left(1 - \frac{4m_f^2}{M_H^2}\right), \tag{2.62}$$

where N_C is the number of colours (1 for leptons and 3 for quarks). The decay width for a generic two-body decay is given by:

$$\Gamma(X \to ij) = \int \frac{1}{32\pi^2} |\mathcal{M}_{X\to ij}|^2 \frac{|\vec{p}_i|}{M_X^2} d\Omega, \tag{2.63}$$

which implies the partial decay width of the Higgs boson to fermions can be written as:

$$\Gamma(H \to f\bar{f}) = N_C \frac{1}{8\pi} \frac{m_f^2}{v^2} M_H \left(1 - \frac{4m_f^2}{M_H^2}\right)^{3/2}. \tag{2.64}$$

The decay widths to weak vector bosons are slightly more complicated since one of the bosons is produced off shell and therefore a three-body decay, where the off-shell boson decays immediately, must be considered. The case of decays to gluons is again more complicated, since is must occur through a loop process. The calculation of these decay widths is not performed here, instead, the final formulae for the various decay channels, calculated with radiative corrections [14, 30], are reported:

Table 2.2 Branching ratios (BR) for the dominant decay channels of a 125 GeV SM Higgs boson [32]

Decay channel	BR [$m_{\mathrm{H}} = 125$ GeV]
$\mathrm{H} \to \mathrm{bb}$	$(5.82 \pm 0.07) \times 10^{-1}$
$\mathrm{H} \to \mathrm{W^+W^-}$	$(2.14 \pm 0.03) \times 10^{-1}$
$\mathrm{H} \to \mathrm{gg}$	$(8.19 \pm 0.42) \times 10^{-2}$
$\mathrm{H} \to \tau^+\tau^-$	$(6.27 \pm 0.10) \times 10^{-2}$
$\mathrm{H} \to \mathrm{cc}$	$(2.89\ ^{+0.16}_{-0.06}) \times 10^{-2}$
$\mathrm{H} \to \mathrm{ZZ}$	$(2.62 \pm 0.04) \times 10^{-2}$
$\mathrm{H} \to \gamma\gamma$	$(2.27 \pm 0.05) \times 10^{-3}$
$\mathrm{H} \to \mathrm{Z}\gamma$	$(1.53 \pm 0.09) \times 10^{-3}$
$\mathrm{H} \to \mu\mu$	$(2.18 \pm 0.04) \times 10^{-6}$

$$\Gamma(H \to \ell^+\ell^-) = \frac{G_{\mathrm{F}} M_H}{4\sqrt{2}\pi} \bar{m}_\ell^2(M_H) \tag{2.65}$$

$$\Gamma(H \to q\bar{q}) = \frac{3 G_{\mathrm{F}} M_H}{4\sqrt{2}\pi} \bar{m}_q^2(M_H) \left[1 + 5.67\frac{\alpha_s}{\pi} + (35.94 - 1.36 N_F)\frac{\alpha_s^2}{\pi^2}\right] \tag{2.66}$$

$$\Gamma(H \to gg) = \frac{G_{\mathrm{F}} \alpha_s^2 M_H^3}{36\sqrt{2}\pi^3} \left[1 + \frac{\alpha_s}{\pi}\left(\frac{95}{4} - \frac{7}{6}N_F + \frac{33 - 2N_F}{6}\log\frac{\mu^2}{M_H^2}\right)\right] \tag{2.67}$$

$$\Gamma(H \to VV^*) = \delta_V' \frac{G_{\mathrm{F}} M_H^3}{16\sqrt{2}\pi} R_T(M_V^2/M_H^2) \tag{2.68}$$

where $\bar{m}_\ell^2(M_H)$ and $\bar{m}_q^2(M_H)$ are the running fermion masses, α_s is defined at the scale M_H, N_F is the number of light-quark flavours, $\mu \sim M_H$, $\delta_W' = 1$, $\delta_Z' = 7/12 - 10\sin^2\theta_W/9 + 40\sin^4\theta_W/9$, and

$$R_T(x) = \frac{3(1 - 8x + 20x^2)}{\sqrt{4x - 1}} \arccos\left(\frac{3x - 1}{2x^{3/2}}\right) - \frac{1 - x}{2x}(2 - 13x + 47x^2) - \frac{3}{2}(1 - 6x + 4x^2)\log x.$$

With the decay widths given in Eqs. (2.65)–(2.68) and others reported in Ref. [14, 31], the total decay width and various branching ratios of the SM Higgs boson can be calculated. After accounting for higher-order corrections, these values are shown as a function of the Higgs boson mass in Fig. 2.5a and 2.5b, respectively. The total width for a 125 GeV Higgs boson is $(4.09 \pm 0.06) \times 10^{-3}$ GeV [32] and the branching ratios for the most dominant channels are listed in Table 2.2. The total width can be used to calculate the lifetime $\tau = \hbar/\Gamma$, and its value is $\tau_{\mathrm{H}} = (1.61 \pm 0.02) \times 10^{-22}$ s.

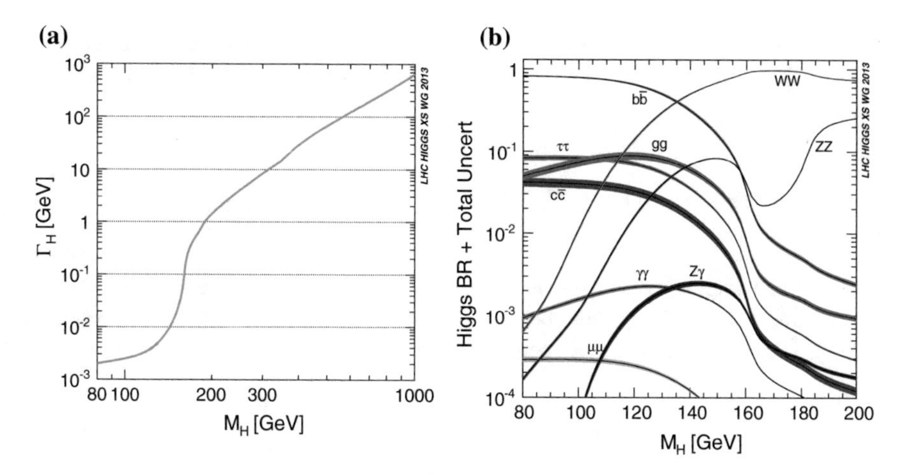

Fig. 2.5 Total decay width Γ_{H} (**a**) and branching ratio (BR) for various decay channels (**b**) of the SM Higgs boson as a function of its mass [32]

2.3.3 Higgs Boson Production

The Higgs boson is not observed in nature as it requires extremely high energies to be produced and it decays almost immediately. The only known way to create the Higgs boson on Earth is in particle colliders, either hadron colliders, such as the LHC and the Tevatron,[4] or at lepton colliders.

The main production mechanisms for the SM Higgs boson depend on the fact that the Higgs boson couplings are proportional to the mass of the coupled particle. Therefore the Higgs boson couples preferentially to heavy particles, i.e. the massive W and Z vector bosons, the top quark and, to a lesser extent, the bottom quark. At a hadron collider this implies that the most dominant production mechanism is gluon–gluon fusion, which proceeds via a loop of heavy quarks, predominantly top quarks and, to a lesser extent, bottom quarks (see Fig. 2.6a).

At LO (with just one loop), the partonic cross section for the gg → H process, using the LO gluonic decay width, cf. Eq. (2.67), is given by [14]:

$$\sigma_{\mathrm{LO}}(\mathrm{gg}\to\mathrm{H}) = \frac{\pi^2}{8M_H^3}\Gamma_{\mathrm{LO}}(\mathrm{H}\to\mathrm{gg})M_H^2\delta(\hat{s}-M_H^2) \tag{2.69}$$

$$= \frac{G_{\mathrm{F}}\alpha_s^2(\mu_{\mathrm{R}}^2)}{288\sqrt{2}\pi}\left|\frac{3}{4}\sum_q A_{1/2}^H(\tau_q)\right|^2 M_H^2\delta(\hat{s}-M_H^2), \tag{2.70}$$

[4]The Tevatron was a circular proton-antiproton accelerator and collider, located at Fermilab in Illinois, US, and operating from 1983 to 2011.

where $\hat{s}$ is the gg invariant energy squared, μ_{R} is the renormalisation scale, $\tau_q = M_H^2/4m_q^2$ is defined by the pole mass m_q of the heavy quark, and the form factor is given by:

$$A_{1/2}^H(\tau) = 2\left[\tau + (\tau - 1)f(\tau)\right]\tau^{-2}, \;\text{where}\; f(\tau) = \begin{cases} \arcsin^2\sqrt{\tau} & \text{if } \tau \leq 1 \\ -\frac{1}{4}\left[\log\frac{1+\sqrt{1-\tau^{-1}}}{1-\sqrt{1-\tau^{-1}}} - i\pi\right]^2 & \text{if } \tau > 1 \end{cases} \tag{2.71}$$

which approaches 4/3 for $\tau \to 0$ ($m_q \gg M_H$) and zero for $\tau \to \infty$ ($m_q \to 0$).

Considering that the gluons come from protons with a parton momentum fraction x, the LO proton–proton cross section can be written as:

$$\sigma_{\mathrm{LO}}(\mathrm{pp} \to \mathrm{H}) = \sigma_{\mathrm{LO}}(\mathrm{gg} \to \mathrm{H})\tau_H \frac{d\mathcal{L}^{\mathrm{gg}}}{d\tau_H}, \tag{2.72}$$

where the gluon luminosity is given by:

$$\frac{d\mathcal{L}^{\mathrm{gg}}}{d\tau} = \int_\tau^1 \frac{dx}{x} g(x, \mu_{\mathrm{F}}^2) g(\tau/x, \mu_{\mathrm{F}}^2), \tag{2.73}$$

and $\tau_H = M_H^2/s$, where s is the square of the collider energy, and g is the parton density function defined at the factorisation scale μ_{F}.

To represent the actual situation in high-energy proton–proton collisions, several corrections need to be made to the LO cross section (2.72), which account for various orders and type (strong or electroweak) of radiation. In the following the main production methods for the Higgs boson at the LHC and their numerically calculated higher-order cross sections are summarised.

Higgs boson production at the LHC

At the LHC, the four main production processes of the Higgs boson, in order of dominance, are:

- gluon–gluon fusion: $\mathrm{gg} \to \mathrm{H}$
- vector boson fusion: $\mathrm{qq} \to \mathrm{qq} + \mathrm{V}^*\mathrm{V}^* \to \mathrm{qq} + \mathrm{H}$
- associated production with W/Z: $\mathrm{q\bar{q}} \to \mathrm{V} + \mathrm{H}$
- associated production with top quarks: $\mathrm{q\bar{q}}, \mathrm{gg} \to \mathrm{t\bar{t}} + \mathrm{H}$

where V is a massive vector boson (W or Z). The LO Feynman diagrams of these four production processes are illustrated in Fig. 2.6. The production cross sections, calculated at various orders with electroweak corrections, as a function of the centre-of-mass energy for a 125 GeV Higgs boson and as a function of m_{H} for $\sqrt{s} = 13$ TeV are shown in Fig. 2.7.

As can be seen from Fig. 2.7, the dominant process is gluon–gluon fusion, with a cross section an order of magnitude greater than vector boson fusion for the entire centre-of-mass energy range. In gluon–gluon fusion, the gluons are indirectly coupled

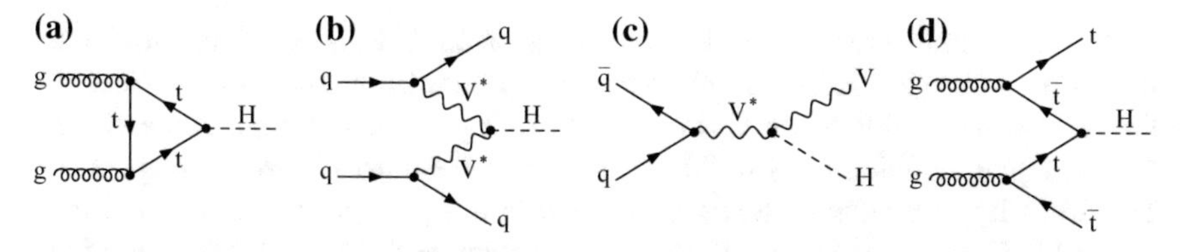

Fig. 2.6 The principal SM Higgs boson production mechanisms at the LHC: (**a**) gluon–gluon fusion; (**b**) vector boson fusion; (**c**) associated production with W/Z (Higgs-strahlung); (**d**) associated production with top quarks

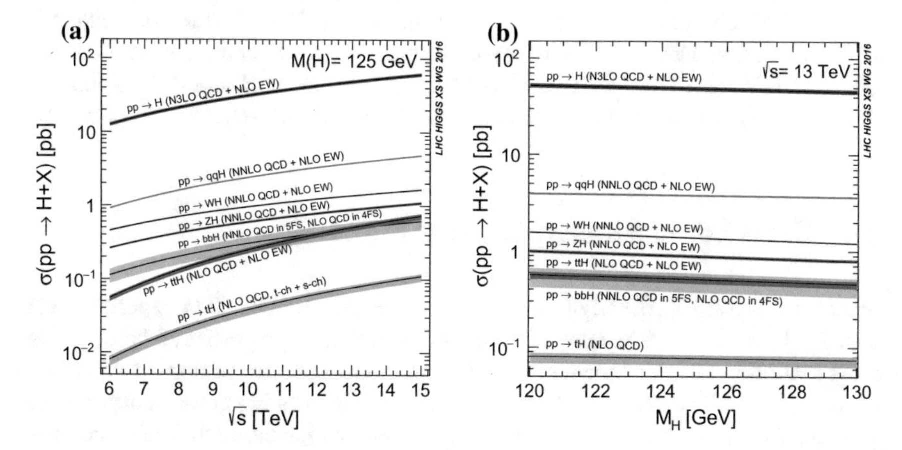

Fig. 2.7 Production cross sections of the Higgs boson at the LHC [32]: (**a**) as a function of the centre-of-mass collision energy for a 125 GeV Higgs boson; (**b**) as a function of the Higgs boson mass for $\sqrt{s} = 13$ TeV. The bands around the curves indicate the theoretical uncertainties related to higher order perturbative corrections

to the Higgs boson via a triangle quark loop, with top quarks being the most dominant, as shown in Fig. 2.6a. There are no other production products in this process. At $m_{\mathrm{H}} = 125$ GeV and $\sqrt{s} = 13$ TeV, the cross section is approximately 49 pb [32].

The next most important production process at the LHC is vector boson fusion (VBF), shown in Fig. 2.6b. At $m_{\mathrm{H}} = 125$ GeV, VBF accounts for about 7% of the total production cross section at 13 TeV, with a cross section of 3.8 pb. In this process the two quarks which radiate the W or Z boson pair continue at small angles to the beam direction in the forward and backward regions. The Higgs boson produced from the fused W or Z pair is usually at wide angles with respect to the beam.

In associated production with a W or Z boson, also known as Higgs-strahlung (see Fig. 2.6c), a quark and antiquark annihilate to form an off-shell W or Z boson which then radiates a Higgs boson and continues as a real boson. Since this process requires an antiquark, which at the LHC must be a sea quark, it has a lower production rate than VBF, despite the same Higgs coupling. The respective cross sections at $\sqrt{s} = 13$ TeV for a 125 GeV Higgs boson produced in association with a W or Z boson are 1.4 pb and 0.88 pb.

The next dominant Higgs production process at the LHC is associated production with top quarks (Fig. 2.6d), which increases with collision energy at a faster rate than the others. In this process, two gluons[5] each produce a $\mathrm{t\bar{t}}$ pair, with a virtual t from one pair annihilating with the virtual $\bar{\mathrm{t}}$ from the other to form a Higgs boson. The difficulty of producing a Higgs boson with a $\mathrm{t\bar{t}}$ pair, given the large top mass ($m_\mathrm{t} \approx 172\,\mathrm{GeV}$), is reflected in the low cross section of this process of about 0.51 pb for $m_\mathrm{H} = 125\,\mathrm{GeV}$ at 13 TeV.

Two other production mechanisms for the Higgs boson at the LHC are associated production with b quarks, $\mathrm{p\bar{p}} \rightarrow \mathrm{b\bar{b}H}$, and single-top quark associated production $\mathrm{p\bar{p}} \rightarrow \mathrm{tH}$, which can occur through the exchange of a W boson in the t or s-channel, or in association with a W boson (tW-channel). The approximate cross sections at $\sqrt{s} = 13\,\mathrm{TeV}$ for a 125 GeV Higgs boson are 0.49 pb for $\mathrm{b\bar{b}H}$ and 0.07, 0.003 and 0.02 pb for tH production in the t-channel, s-channel and tW-channel, respectively.

2.4 Higgs Boson Measurements at the LHC

Since the discovery of the Higgs boson in 2012, the ATLAS and CMS experiments at the LHC have performed a number of measurements of its properties and production cross section in different channels. In 2011 the LHC operated proton–proton collisions at $\sqrt{s} = 7\,\mathrm{TeV}$ and provided data corresponding to an integrated luminosity of around $5\,\mathrm{fb}^{-1}$ to each experiment. In 2012, the energy increased to 8 TeV and integrated luminosity also increased to around $20\,\mathrm{fb}^{-1}$ per experiment. Together, these two data-taking years are referred to as Run 1. With the Run 1 data, ATLAS and CMS were able to observe the Higgs boson in several production and decay channels, thus providing estimates of its cross section, decay rates and couplings [33], as well as measure its mass [34].

The production and decays rates are measured in terms of a signal strength μ, which is the ratio of the measured production cross section or decay branching ratio to the SM prediction:

$$\mu = \sigma/\sigma_\mathrm{SM} \quad \text{or} \quad \mu = \mathrm{BR}/\mathrm{BR}_\mathrm{SM} \tag{2.74}$$

Similarly a coupling modifier κ, defined as the square root of the ratio of the measured cross section or decay width to the SM prediction:

$$\kappa^2 = \sigma/\sigma_\mathrm{SM} \quad \text{or} \quad \kappa^2 = \Gamma/\Gamma_\mathrm{SM}, \tag{2.75}$$

is used to measure the Higgs boson couplings to bosons and fermions.

From the Run 1 results, the combined measurement of the Higgs boson mass is $125.09 \pm 0.24\,\mathrm{GeV}$, as shown in Fig. 2.8, and all measured couplings are consistent

[5] The process is also initiated by quarks, although with a rate about 1/5 that for the gluon initiated subprocess.

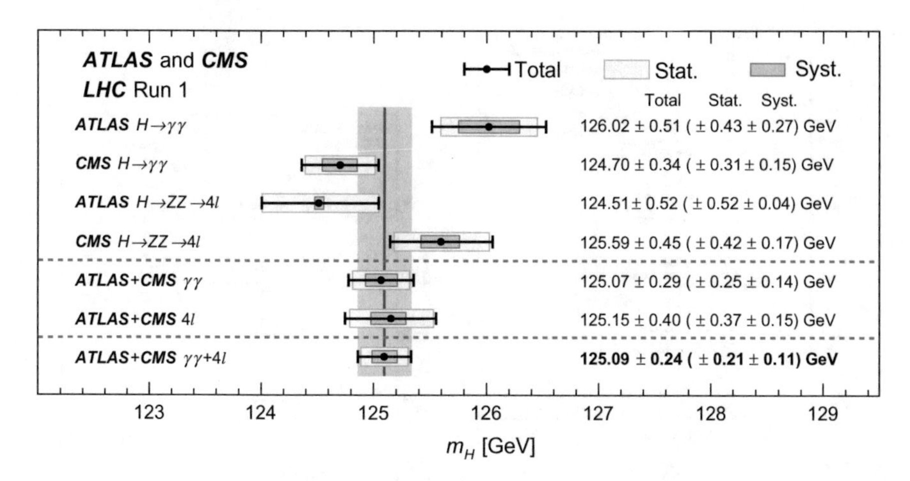

Fig. 2.8 Summary of Higgs boson mass measurements from the individual analyses of ATLAS and CMS and their combination [34]. The systematic (narrower bands), statistical (wider bands), and total (black error bars) uncertainties are indicated. The vertical line and corresponding shaded column indicate the central value and the total uncertainty of the combined measurement, respectively

with their SM values, as shown in Fig. 2.9. The measurement of the coupling modifier κ_t includes direct $t\bar{t}H$ production, but also contributions from indirect processes involving top-quark loops, such as gluon–gluon fusion production and decays to photons. It is these indirect processes which drive the precision on the measurement of κ_t. Despite the high precision obtained on the measurements of the properties of the Higgs sector, they could all benefit from additional data.

In 2015 and 2016, the LHC provided pp collisions at 13 TeV, corresponding to integrated luminosities per experiment of approximately 3 and 35 fb^{-1} respectively. A number of individual search results and some combined results have already been produced, with a full combination of 13 TeV data across all production and decay channels recently performed by CMS [35]. This combination measures the top quark coupling modifier to be $\kappa_t = 1.11^{+0.12}_{-0.11}$. The latest results from both experiments grouped by search signature are summarised below.

2.4.1 Decays to Vector Bosons

The decays to ZZ, where each Z boson decays to two leptons ($H \rightarrow ZZ^* \rightarrow 4\ell$), and to $\gamma\gamma$ provide very clean, fully reconstructed resonant-mass peaks, which can provide precision measurements of the Higgs boson mass. The $H \rightarrow ZZ^* \rightarrow 4\ell$ decay provides few signal events on a very small background, as can be seen in Fig. 2.10 for CMS and ATLAS. The $H \rightarrow \gamma\gamma$ channel provides a relatively large number of events on a significantly larger, yet well understood background, from

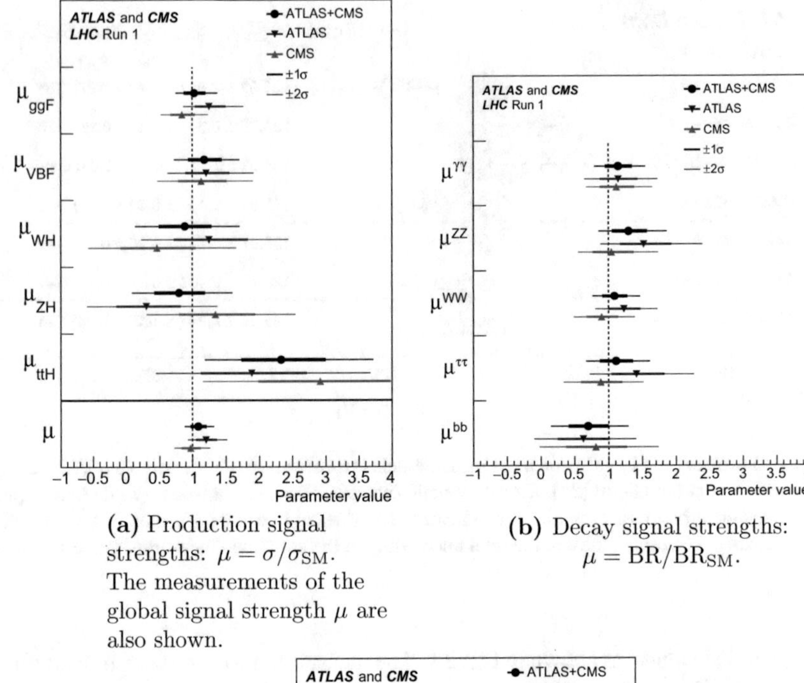

(a) Production signal strengths: $\mu = \sigma/\sigma_{\mathrm{SM}}$. The measurements of the global signal strength μ are also shown.

(b) Decay signal strengths: $\mu = \mathrm{BR}/\mathrm{BR}_{\mathrm{SM}}$.

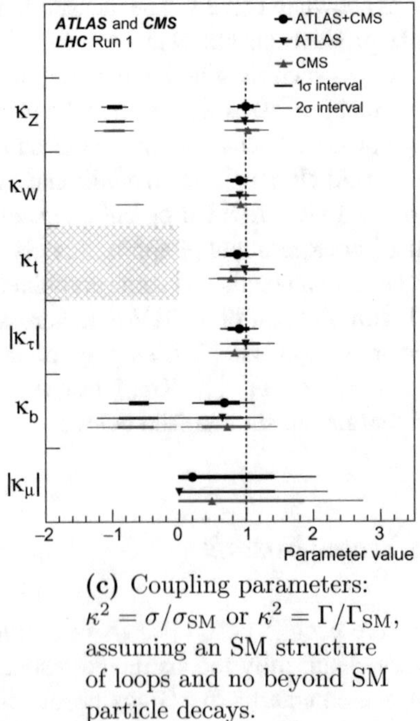

(c) Coupling parameters: $\kappa^2 = \sigma/\sigma_{\mathrm{SM}}$ or $\kappa^2 = \Gamma/\Gamma_{\mathrm{SM}}$, assuming an SM structure of loops and no beyond SM particle decays.

Fig. 2.9 Best fit values of various parameters for the combination of ATLAS and CMS data, and separately for each experiment [33]. The error bars indicate the 1σ (thick lines) and 2σ (thin lines) intervals

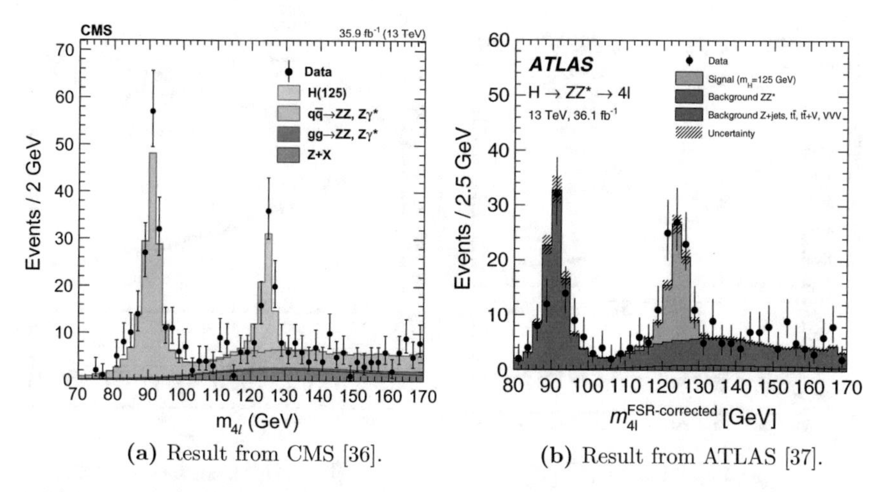

(a) Result from CMS [36]. (b) Result from ATLAS [37].

Fig. 2.10 Four lepton (4μ, $2e2\mu$, 4e) invariant mass distribution in the $H \rightarrow ZZ \rightarrow 4\ell$ decay channel

which a clear peak can be extracted, as shown in Fig. 2.11 for CMS and ATLAS. Each experiment also provides a breakdown of the cross section by production process, which are compatible with the SM, as seen in Fig. 2.12. The best estimates of the Higgs boson mass for CMS and ATLAS are $m_{\mathrm{H}} = 125.26 \pm 0.21$ GeV [36] and $m_{\mathrm{H}} = 124.97 \pm 0.28$ GeV [37], respectively.

The decays to W^+W^- are more complicated to measure than those to ZZ and $\gamma\gamma$, since the leptonic decay of the W boson includes neutrinos and thus the event cannot be fully reconstructed. However, the channel does benefit from a relatively high branching ratio (cf. Table 2.2). The approach adopted is to search for oppositely charged electron-muon pairs ($\mathrm{H} \rightarrow W^+W^- \rightarrow e^+\nu_e\mu^-\bar{\nu}_\mu/\mu^+\nu_\mu e^-\bar{\nu}_e$) and use a combination of the dilepton invariant mass and the Higgs transverse mass (calculated from the transverse momentum of the leptons and the missing transverse momentum) to extract the signal. The results at 13 TeV are presented in terms of signal significance and the signal strength relative to the SM prediction. For CMS, with 35.9 fb^{-1} of 13 TeV data, the observed and expected significance is 9.1 and 7.1 standard deviations, respectively, while the best fit value of the cross section times branching ratio relative to the SM prediction is $1.28^{+0.18}_{-0.17}$, which is dominated by systematic uncertainties [38]. For ATLAS, the latest results are with 36.1 fb^{-1} of 13 TeV data and correspond to respective observed and expected significances of 6.3 and 5.2 standard deviations, with a best fit signal strength of $1.21^{+0.22}_{-0.21}$ [39].

2.4.2 Decays to Fermions

Results in the $\mathrm{H} \rightarrow \tau^+\tau^-$ decay channel with 35.9 fb^{-1} of 13 TeV data have been produced by CMS targeting the gluon fusion and VBF production modes. The search

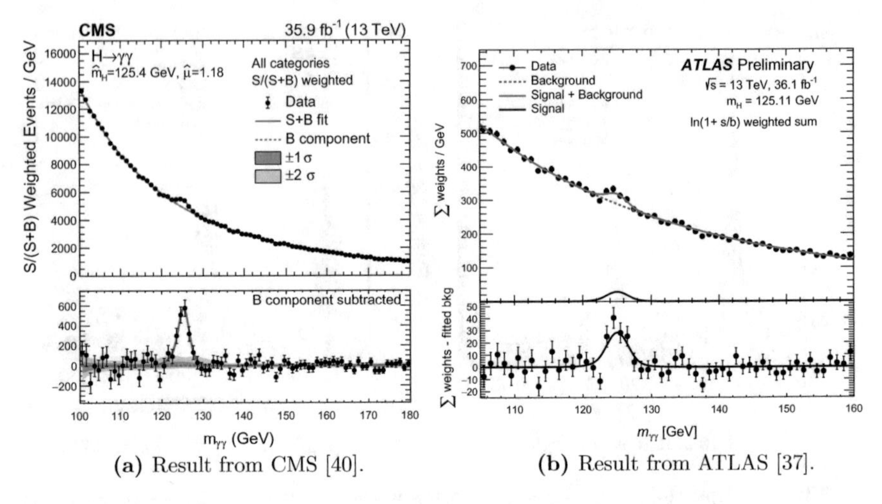

(a) Result from CMS [40]. (b) Result from ATLAS [37].

Fig. 2.11 Diphoton invariant mass distribution in the $\mathrm{H} \rightarrow \gamma\gamma$ decay channel

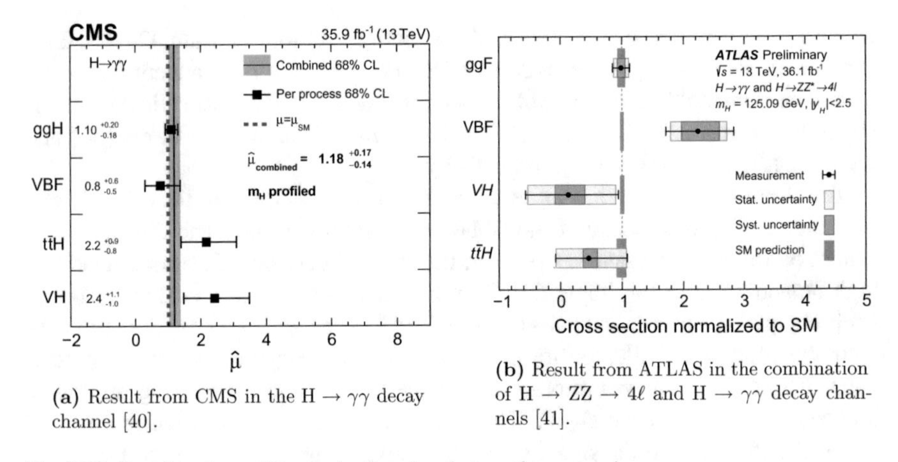

(a) Result from CMS in the $\mathrm{H} \rightarrow \gamma\gamma$ decay channel [40].

(b) Result from ATLAS in the combination of $\mathrm{H} \rightarrow \mathrm{ZZ} \rightarrow 4\ell$ and $\mathrm{H} \rightarrow \gamma\gamma$ decay channels [41].

Fig. 2.12 Best fit values of the production signal strengths: $\mu = \sigma/\sigma_{\mathrm{SM}}$

selects events with both hadronic and leptonic decays of the τ lepton, leading to four different decay channels: $\tau_\mathrm{h}\tau_\mathrm{h}$, $\mathrm{e}\tau_\mathrm{h}$, $\mu\tau_\mathrm{h}$, and $\mathrm{e}\mu$, where τ_h is a hadronically decaying τ jet. Given the presence of neutrinos in all these decay channels, a clear reconstruction of the $\tau\tau$ invariant mass cannot be performed. Instead, a likelihood method is used to estimate $m_{\tau\tau}$ in some cases and the visible decay products are used to reconstruct a visible $\tau\tau$ mass in others. The results correspond to an observed and expected significance of 4.9 and 4.7 standard deviations, respectively, and a best fit signal strength, $\sigma/\sigma_{\mathrm{SM}}$, of $1.06^{+0.25}_{-0.24}$ [42]. When combined with the data collected at 7 and 8 TeV, this leads to an observed significance of 5.9 standard deviations, which is equal to the expected significance and represents the first observation of Higgs boson decays to τ leptons by a single experiment.

The search for the Higgs boson decaying into a pair of bottom quarks is performed separately in three production channels, namely VBF, associated production with a vector boson, and $t\bar{t}H$. The search is additionally performed in an inclusive boosted region which includes the gluon fusion production mode [43]. A resolved (non-boosted) search in the gluon fusion production channel is not performed due to the overwhelming background such a signature would entail. Two of the resolved searches are summarised below, while the search in the $t\bar{t}H$ production channel is described in Sect. 2.4.3.

Evidence for the $H \rightarrow b\bar{b}$ decay has been presented in associated production with W/Z at 13 TeV by both ATLAS [44] and CMS [45]. The searches focus on the leptonic decays of the weak vector boson and reconstruct the Higgs boson by selecting two b jets. The ultimate discriminating variable is a multivariate discriminant based on event information primarily about these two jets, leptons and missing transverse momentum. With 36.1 fb^{-1}, ATLAS published an observed and expected significance of 3.5 and 3.0 standard deviations, respectively, corresponding to a best fit signal strength of $\mu = 1.20^{+0.42}_{-0.36}$, which is dominated by systematic uncertainties. CMS published a similar result using 35.9 fb^{-1} of data with respective observed and expected significances of 3.3 and 2.8 standard deviations, and a best fit μ of $1.19^{+0.40}_{-0.38}$, also dominated by systematic uncertainties.

The search for the $H \rightarrow b\bar{b}$ decay in the vector boson fusion production channel has been performed by ATLAS, with 12.6 fb^{-1} of 13 TeV data [46] and CMS, with 2.3 fb^{-1} of 2015 data only [47]. The production and decay signature involves four jets, two of which are b jets, and thus suffers from a significant amount of background from QCD interactions. To reduce the background and have a purer signal, the search performed by ATLAS focuses on production in association with a photon, thus including a photon in their event selection. ATLAS reported a best fit value of the signal strength of $\mu = \sigma/\sigma_{\text{SM}} = -3.9^{+2.8}_{-2.7}$, while the best fit CMS result is $\mu = -3.7^{+2.4}_{-2.5}$.

The search for $H \rightarrow \mu^+\mu^-$ decays at 13 TeV has been presented separately by ATLAS and CMS. It involves a rare decay process of the Higgs boson and selects events with two opposite-charge muons to recreate a dimuon invariant mass. Because of the low branching ratio of this decay and the large background from the $Z/\gamma^* \rightarrow \mu\mu$ process, the search has a low sensitivity. With 36.1 fb^{-1} of 13 TeV data, ATLAS found no excess of events, with a best fit signal strength of $\mu = -0.1 \pm 1.5$, dominated by statistical uncertainties, and observed and expected 95% confidence level upper limits on μ of 3.0 and 3.1, respectively [48]. With 35.9 fb^{-1} of data, CMS obtained a best fit signal strength of $\mu = 0.7 \pm 1.0$ and observed and expected upper limits of $\mu < 2.64$ and 2.08, respectively [49]. When combined with the CMS results from Run 1, the best fit signal strength is $\mu = 0.9^{+1.0}_{-0.9}$ and the respective observed and expected upper limits are 2.64 and 1.89 times the SM value.

2.4.3 Associated Production with Top Quarks

Because of the complicated final state with many jets, the search for $\mathrm{t\bar{t}H}$ production is performed in isolation for the $\mathrm{H} \rightarrow \mathrm{b\bar{b}}$, $\mathrm{H} \rightarrow \tau^+\tau^-$ and $\mathrm{H} \rightarrow \mathrm{W^+W^-}$ decay channels, while it is performed inclusively in the $\mathrm{H} \rightarrow \mathrm{ZZ} \rightarrow 4\ell$ and $\mathrm{H} \rightarrow \gamma\gamma$ decay channels, as shown in Sect. 2.4.1.

The $\mathrm{H} \rightarrow \mathrm{W^+W^-}$, $\mathrm{H} \rightarrow \tau^+\tau^-$, and $\mathrm{H} \rightarrow \mathrm{ZZ} \not\rightarrow 4\ell$ decay modes are all covered in a search for the Higgs boson in final states with multiple leptons, referred to as *multilepton* final states. This search selects events with two same sign electrons or muons, or three or more leptons including at least one electron or muon and up to two hadronically-decaying τ leptons. A multivariate analysis discriminant is built from the event information and used to extract the signal. With $36.1\,\mathrm{fb}^{-1}$ of 13 TeV data, ATLAS observed a significance of 4.1 standard deviations (expected 2.8), and a best fit signal strength for the $\mathrm{t\bar{t}H}$ cross section relative to the SM prediction of $\mu = 1.6^{+0.5}_{-0.4}$, which is dominated by systematic uncertainties [50]. On the other hand, CMS analysed $35.9\,\mathrm{fb}^{-1}$ of 13 TeV data and measured a signal strength $1.2^{+0.5}_{-0.4}$ times the SM cross section, with an observed and expected significance of 3.2 and 2.8 standard deviations, respectively [51].

The $\mathrm{H} \rightarrow \mathrm{b\bar{b}}$ decay modes of $\mathrm{t\bar{t}H}$ production are covered by two separate searches at CMS and ATLAS: one including leptonic decays of the top quarks, and the other selecting all-jet final states targeting hadronic decays of the top quarks. The fully hadronic search has been completed by ATLAS at 8 TeV, while the CMS search at 13 TeV is the subject of this thesis. The leptonic search includes final states with one or two electrons or muons and four or six jets, of which four are b jets. The complicated final state with uncertain jet-to-quark matching and missing energy requires a multivariate discriminant to separate the signal from the relatively large background, dominated by $\mathrm{t\bar{t}}$ + jets production. ATLAS has presented results using $36.1\,\mathrm{fb}^{-1}$ of 13 TeV data, with a best fit signal strength of $\mu = 0.84^{+0.64}_{-0.61}$ and observed and expected significances over the background only hypothesis of 1.4 and 1.6 standard deviations, respectively [52]. The equivalent results for CMS are based on $35.9\,\mathrm{fb}^{-1}$ of 13 TeV data and lead to a best fit signal strength of $\mu = 0.72 \pm 0.45$, with observed and expected significances of 1.6 and 2.2 standard deviations, respectively [53]. Both of these searches are dominated by systematic uncertainties.

In addition to the individual $\mathrm{t\bar{t}H}$ searches described above, ATLAS also performed a combination of the $\mathrm{t\bar{t}H}$ production component of all decay channels, namely $\mathrm{H} \rightarrow \mathrm{ZZ} \rightarrow 4\ell$, $\mathrm{H} \rightarrow \gamma\gamma$, multilepton and $\mathrm{H} \rightarrow \mathrm{b\bar{b}}$ [50]. The results of this combination culminated in an observed and expected significance of 4.2 and 3.8 standard deviations, respectively, with a best fit value for the $\mathrm{t\bar{t}H}$ signal strength of $\mu = 1.17^{+0.33}_{-0.30}$, which is dominated by systematic uncertainties. This constituted the strongest evidence for $\mathrm{t\bar{t}H}$ production by a single experiment up until the publication of the analysis underlying this thesis and the subsequent combination by CMS. The latter results in the first ever observation of $\mathrm{t\bar{t}H}$ production with observed and expected significances of 5.2 and 4.2 standard deviations, respectively, and a best fit signal strength of $\mu = 1.26^{+0.31}_{-0.26}$ [54].

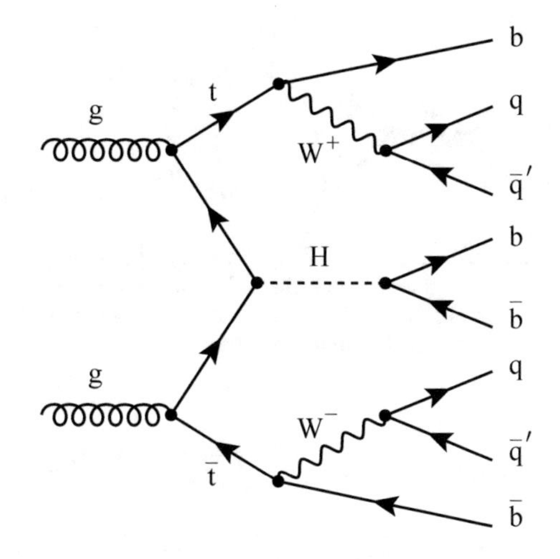

Fig. 2.13 LO Feynman diagram for the most common $t\bar{t}H$ production process at the LHC, including the subsequent hadronic decays of the top quark-antiquark pair as well as the decay of the Higgs boson into a bottom quark-antiquark pair

2.5 Fully Hadronic $t\bar{t}H$ ($H \rightarrow b\bar{b}$) Channel

The subject of this thesis is the search for the Higgs boson in the $t\bar{t}H$ production channel and the $H \rightarrow b\bar{b}$ decay channel where the W bosons from both top quarks decay to light quarks. The signature of this signal thus contains eight final-state jets, four of which are b jets. The LO Feynman diagram representing this process, when initiated by gluons, is shown in Fig. 2.13.[6] The final decay products are eight quarks which all hadronise into jets, typically produced at large angles with respect to the beam axis and thus having a relatively high transverse momentum.

2.5.1 *Theoretical Cross Section*

The complete analytical expression for the LO $gg \rightarrow t\bar{t}H$ cross section is too complicated to derive explicitly here. It must consider the processes illustrated in Fig. 2.14b–d, plus all unique vertex permutations of exchanging the fermion with the antifermion and the gluons with each other—a total of 8 diagrams. Nevertheless, the basic steps to its calculation are outlined in the following.

Following the notation of Ref. [14], we begin by denoting the four-momenta of the incoming gluons, top quark, top antiquark and Higgs boson respectively by g_1, g_2, p, $\bar{p}$ and k, and the gluon polarisation four-vectors as ϵ_1 and ϵ_2. The invariant mass squared of the initial gluons is given by $\hat{s} = Q^2 = (g_1 + g_2)^2 = (p + \bar{p} + k)^2$ and the LO scattering amplitudes for the three diagrams shown in Figs. 2.14b–d, labelled $\mathcal{M}_1$, $\mathcal{M}_2$ and $\mathcal{M}_3$, respectively, are given by [55]:

[6]The Higgs boson can also be emitted from external top quark lines, in which case the process can be initiated by quark-antiquark annihilation (Fig. 2.14a) or gluon fusion (Figs. 2.14b, d).

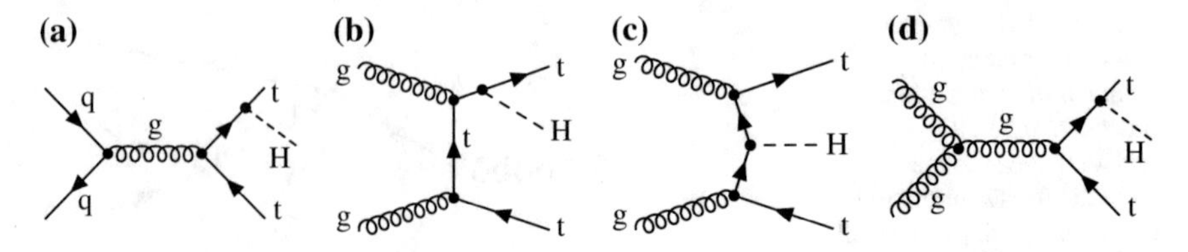

Fig. 2.14 Examples of LO Feynman diagrams for tt̄H production: (**a**) initiated by quarks;(**b**) initiated by gluons with t-channel exchange and radiation from external lines (**c**) initiated by gluons with t-channel exchange and radiation from internal lines; (**d**) initiated by gluons with s-channel exchange and radiation from external lines

$$
\begin{aligned}
\mathcal{M}_1 &= -AX^a_{ik}X^b_{kj}\bar{u}^j(p)\frac{\not{k}+\not{p}+m_t}{2p\cdot k+M_H^2}\not{\epsilon}_2\frac{-\not{\bar{p}}+\not{g}_1+m_t}{-2g_1\cdot\bar{p}}\not{\epsilon}_1 v^i(\bar{p}) + \left\{\begin{matrix} g_1 \leftrightarrow g_2, \epsilon_1 \leftrightarrow \epsilon_2 \\ g_1 \leftrightarrow g_2, \epsilon_1 \leftrightarrow \epsilon_2, p \leftrightarrow \bar{p} \\ p \leftrightarrow \bar{p} \end{matrix}\right\} \\
\mathcal{M}_2 &= -AX^a_{ik}X^b_{kj}\bar{u}^j(p)\not{\epsilon}_2\frac{\not{p}-\not{g}_2+m_t}{-p\cdot g_2}\frac{-\not{\bar{p}}+\not{g}_1+m_t}{-g_1\cdot\bar{p}}\not{\epsilon}_1 v^i(\bar{p}) + \{g_1 \leftrightarrow g_2, \epsilon_1 \leftrightarrow \epsilon_2\} \\
\mathcal{M}_3 &= iAf^{abc}X^c_{ij}\bar{u}^j(p)\frac{\not{\epsilon}_1\not{\epsilon}_2 Q^\lambda}{\hat{s}}\left[2g_1^\nu g^{\lambda\mu} + (g_2-g_1)^\lambda g^{\mu\nu} - 2g_2^\mu g^{\nu\lambda}\right]\frac{\not{\bar{p}}+\not{k}-m_t}{2k\cdot\bar{p}+M_H^2}v^i(\bar{p}) + \{p \leftrightarrow \bar{p}\}
\end{aligned}
\tag{2.76}
$$

where $A = 4\pi\alpha_s(\sqrt{2}m_t^2 G_\mathrm{F})^{1/2}$ are the coupling factors, and the SU(3) generators X^a and structure constants f^{abc} are the same as those in Sect. 2.1.1. The polarisation vectors obey the transversality condition $\epsilon_i \cdot g_i = 0$ and the SU(3) gauge invariance implies $\epsilon_1 \cdot g_2 = \epsilon_2 \cdot g_1$ and invariance under the substitutions $\epsilon_i \leftrightarrow g_i$.

The amplitude squared needs to be summed over the colour and spin states of the final quarks, and averaged over the colour and polarisation states of the initial gluons:

$$|\mathcal{M}|^2 = \frac{1}{256}\sum_{\text{spin,col}}|\mathcal{M}_1 + \mathcal{M}_2 + \mathcal{M}_3|^2. \tag{2.77}$$

The trace over the γ matrices and sum over the indices of the generators and structure function yields:

$$(X^a_{ik}X^b_{kj})^2 = 24, \quad (f^{abc}X^c_{ij})^2 = 12, \quad (X^a_{ik}X^b_{kj})(f^{abc}X^c_{ij}) = 0, \tag{2.78}$$

while the average over the gluon polarisation states must be performed in an axial gauge (since the gluons are massless), for example:

$$\sum_{\lambda_i=1}^{2}\epsilon_i^\mu(g_i,\lambda_i)\epsilon_i^\nu(g_i,\lambda_i) = -g^{\mu\nu} + \frac{2}{\hat{s}}(g_1^\mu g_2^\nu + g_1^\nu g_2^\mu). \tag{2.79}$$

The resulting expression for the amplitude squared is too long to reproduce here.

The cross section for the core gg → tt̄H process is then obtained by integrating over the phase space as:

$$\hat{\sigma}_{\mathrm{LO}} = \frac{1}{\hat{s}} \frac{\alpha_s^2 G_{\mathrm{F}} m_t^2}{\sqrt{2}\pi^3 (2\pi)^9} \int \frac{\mathrm{d}^3 p}{2E_{\mathrm{t}}} \frac{\mathrm{d}^3 \bar{p}}{2E_{\bar{\mathrm{t}}}} \frac{\mathrm{d}^3 k}{2E_{\mathrm{H}}} \delta^{(4)}(Q - p - \bar{p} - k) \, |\mathcal{M}|^2 . \tag{2.80}$$

This parton level cross section must then be folded with the gluon luminosity, cf. Eq. (2.73), to obtain the full cross section for the process $\mathrm{pp} \to \mathrm{gg} \to \mathrm{t\bar{t}H}$:

$$\sigma_{\mathrm{LO}} = \int_0^1 \frac{1}{2} \Big[g(x_1, \mu_{\mathrm{F}}) g(x_2, \mu_{\mathrm{F}}) \hat{\sigma}_{\mathrm{LO}}(x_1, x_2, \mu_{\mathrm{F}}) + \{x_1 \leftrightarrow x_2\} \Big] \mathrm{d}x_1 \mathrm{d}x_2 . \tag{2.81}$$

At this stage it remains to add the top quark and Higgs boson decays. The scattering amplitude (2.77) must be multiplied by the decay amplitudes to give:

$$\left|\mathcal{M}_{\mathrm{gg \to t\bar{t}H \to qqb,qqb,bb}}\right|^2 = |\mathcal{M}|^2 \cdot \left|\mathcal{M}_{\mathrm{t \to qqb}}\right|^2 \cdot \left|\mathcal{M}_{\mathrm{\bar{t} \to qqb}}\right|^2 \cdot \left|\mathcal{M}_{\mathrm{H \to b\bar{b}}}\right|^2 . \tag{2.82}$$

The top quark and Higgs boson decay amplitudes can be simplified with the narrow-width approximation and expressed in terms of the vertex amplitudes:

$$\begin{aligned} \left|\mathcal{M}_{\mathrm{t \to qqb}}\right|^2 &= \frac{\pi}{m_t \Gamma_t} \delta(p^2 - m_t^2) \left|\mathcal{M}_{\mathrm{q,q,b}}\right|^2 \\ \left|\mathcal{M}_{\mathrm{H \to b\bar{b}}}\right|^2 &= \frac{\pi}{m_H \Gamma_H} \delta(k^2 - m_H^2) \left|\mathcal{M}_{\mathrm{b,b}}\right|^2 . \end{aligned} \tag{2.83}$$

The phase space must now only include the final state quarks. Denoting the four-momenta of the top quark decay products as q_1, q_1', b_1, those of the top antiquark as q_2, q_2', b_2 and those of the Higgs boson as $b, \bar{b}$, the phase space volume is parameterised as:

$$\mathrm{d}\Phi = \frac{1}{(2\pi)^{24}} \frac{\mathrm{d}\vec{q}_1}{2E_{q_1}} \frac{\mathrm{d}\vec{q}_1'}{2E_{q_1'}} \frac{\mathrm{d}\vec{b}_1}{2E_{b_1}} \frac{\mathrm{d}\vec{q}_2}{2E_{q_2}} \frac{\mathrm{d}\vec{q}_2'}{2E_{q_2'}} \frac{\mathrm{d}\vec{b}_2}{2E_{b_2}} \frac{\mathrm{d}\vec{b}}{2E_b} \frac{\mathrm{d}\vec{b}}{2E_{\bar{b}}} . \tag{2.84}$$

The cross section for the gluon initiated $\mathrm{t\bar{t}H}$ process in the fully hadronic decay channel is therefore given by:

$$\sigma_{\mathrm{LO}}^{\mathrm{gg \to t\bar{t}H \to 8}q} = \frac{1}{\hat{s}} \frac{\alpha_s^2 G_{\mathrm{F}} m_t^2}{\sqrt{2}\pi^3} \int \mathrm{d}\Phi \delta^{(4)}(Q - \sum_{i=1}^{8} p_i) \left|\mathcal{M}_{\mathrm{gg \to t\bar{t}H \to qqb,qqb,bb}}\right|^2 , \tag{2.85}$$

and the final cross section starting from protons is expressed as:

$$\sigma_{\mathrm{LO}}^{\mathrm{pp \to t\bar{t}H \to 8}q} = \int_0^1 \frac{1}{2} \Big[g(x_1, \mu_{\mathrm{F}}) g(x_2, \mu_{\mathrm{F}}) \sigma_{\mathrm{LO}}^{\mathrm{gg \to t\bar{t}H \to 8}q} + \{x_1 \leftrightarrow x_2\} \Big] \mathrm{d}x_1 \mathrm{d}x_2 . \tag{2.86}$$

This LO cross section will be revisited again as a core component of the analysis strategy described in Chap. 5.

2.5.2 Standard Model Backgrounds

There are several SM processes that can produce the same final state as the fully hadronic $\mathrm{t\bar{t}H}$ signal, with eight jets including four b jets. The underlying production mechanisms vary substantially, but in all cases the required number of jets is reached only through radiation. Nevertheless, in high-energy proton–proton collisions QCD radiation is very common, even up to several consecutive splittings, thus ensuring that the signal rate is overwhelmed by SM background. Furthermore, the presence of four real b jets is not necessary for background processes as there is a significant probability of one or more light-flavour jets to be incorrectly identified as a b jet in the detector (see Sect. 4.3.6).

The SM backgrounds and their main features are described below in order of dominance, while example Feynman diagrams representing some possible processes are illustrated in Fig. 2.15.

- QCD multijet: By far the most dominant background is from jets produced through the strong interaction, referred to as QCD multijet events. Such events include multiple gluon radiation and have a large cross section which drops off as the jet and b jet multiplicity increase and the jet p_{T} increase. Nevertheless, at eight jets with high p_{T} the cross section is still substantially above the signal. Some examples of possible processes are shown in Fig. 2.15a, b.
- $\mathrm{t\bar{t}}$ + jets: The SM $\mathrm{t\bar{t}}$ production with additional jets from radiation forms a large and difficult background, as it has a large cross section and involves a final state with very similar kinematic properties to the signal. An example Feynman diagram for this process is given in Fig. 2.15c. This process is considered irreducible when the additional jets are b jets, and is then referred to as $\mathrm{t\bar{t}} + \mathrm{b\bar{b}}$, with a Feynman diagram shown in Fig. 2.15d. If the additional jets are c jets, as shown in Fig. 2.15e, there is a larger probability of misidentifying them as b jets, making the process more difficult to distinguish from the signal.
- Single top quark: Single top quark production (single t) constitutes the next most dominant background, although it is considered a minor background. It has a larger cross section than the signal, but since it requires many additional radiated jets, its total contribution in the selected final state is less than the signal. The process can occur through an exchange of a W boson in the t or s-channel or in the tW-channel, as shown in Fig. 2.15f, g and h, respectively.
- W + jets: W boson production has a much larger cross section than the signal, however to form a background it requires a significant amount of radiation, which effectively reduces its cross section to below that of the signal. A typical Feynman diagram for this process is given in Fig. 2.15i.
- Z + jets: Z boson production has a lower cross section than W boson production, and at the jet and b jet multiplicity of the signal, it also has a lower cross section than W + jets. The production process is illustrated in Fig. 2.15j.
- $\mathrm{t\bar{t}}$ + Z: $\mathrm{t\bar{t}}$ production in association with a Z boson has a similar cross section to $\mathrm{t\bar{t}H}$ production, however the branching ratio for $\mathrm{Z} \rightarrow \mathrm{b\bar{b}}$ is lower than that for

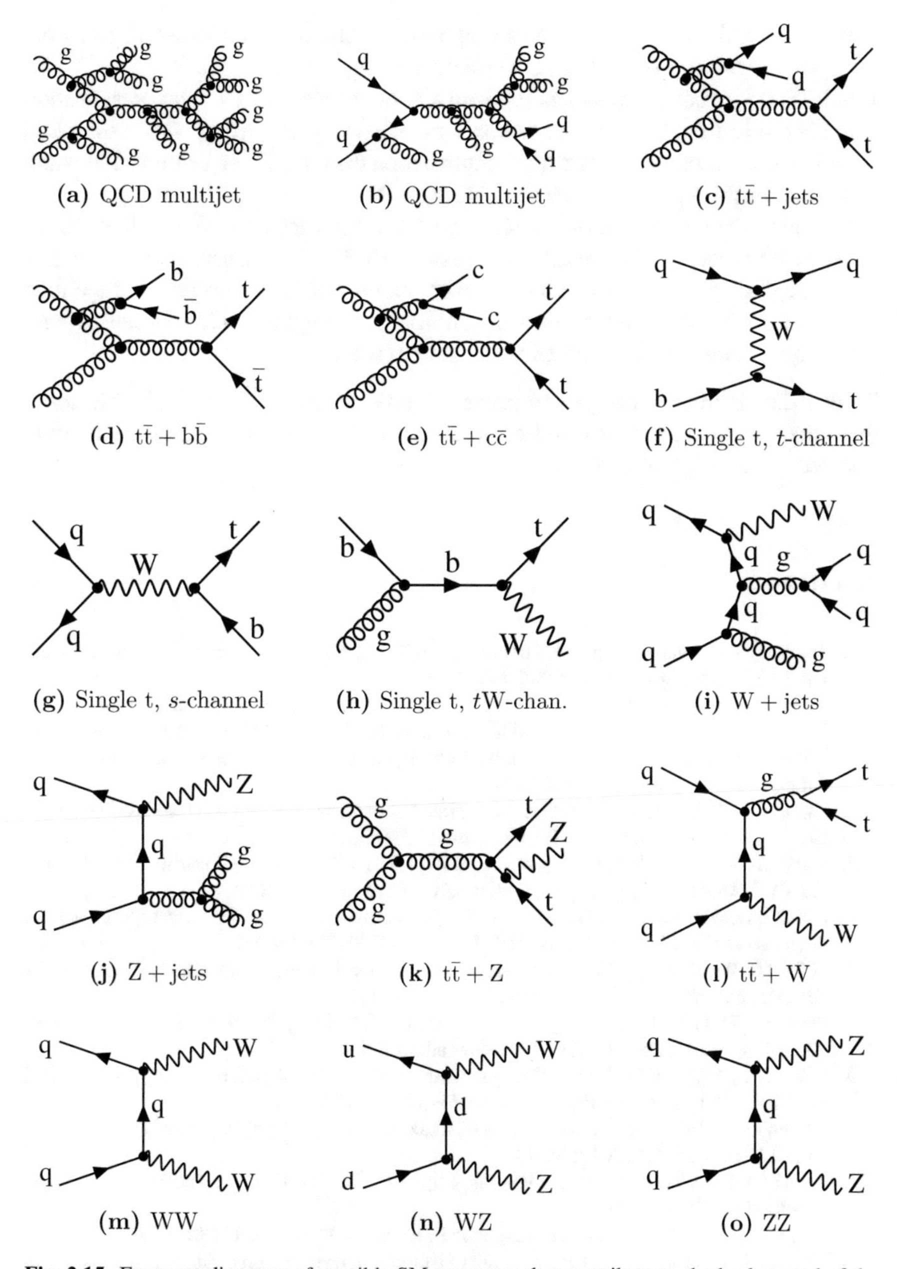

(a) QCD multijet (b) QCD multijet (c) $t\bar{t}$ + jets

(d) $t\bar{t} + b\bar{b}$ (e) $t\bar{t} + c\bar{c}$ (f) Single t, t-channel

(g) Single t, s-channel (h) Single t, tW-chan. (i) W + jets

(j) Z + jets (k) $t\bar{t}$ + Z (l) $t\bar{t}$ + W

(m) WW (n) WZ (o) ZZ

Fig. 2.15 Feynman diagrams of possible SM processes that contribute to the background of the fully hadronic $t\bar{t}H$, $H \to b\bar{b}$ signal. Additional radiation that increases the number of final-state jets is shown for some processes, namely QCD multijet, $t\bar{t}$ + jets, $t\bar{t} + b\bar{b}$, $t\bar{t} + c\bar{c}$, W + jets and Z + jets

$H \rightarrow b\bar{b}$, and therefore it presents a signal-like final state at a lower rate than the signal. A typical process diagram is shown in Fig. 2.15k.

- $t\bar{t} + W$: $t\bar{t}$ production in association with a W boson also has a similar cross section to $t\bar{t}H$ production, however the W boson cannot decay to two b quarks. It therefore makes an even smaller background contribution than $t\bar{t} + Z$. An example Feynman digram for this process can be seen in Fig. 2.15l.
- Diboson: The production of two weak vector bosons occurs as WW, WZ or ZZ in decreasing order of cross section, and is shown in Fig. 2.15m, n and o, respectively. Although the three processes have a cross section one to two orders of magnitude larger than the signal, the number of additional jets required to form a background is large and thus the final contribution is very small.

Further details of the background processes considered in this analysis, including their cross sections, are provided in Sect. 6.1, while their contribution to the final selected events is given in Sect. 6.3.2.

References

1. Thackray AW (1966) The origin of dalton's chemical atomic theory: daltonian doubts resolved. Isis 57:35. https://doi.org/10.1086/350077
2. Rutherford E (1911) The scattering of alpha and beta particles by matter and the structure of the atom. Phil Mag Ser 6(21):669–688. https://doi.org/10.1080/14786440508637080
3. Gell-Mann M (1964) A schematic model of baryons and mesons. Phys Lett 8:214. https://doi.org/10.1016/S0031-9163(64)92001-3
4. Zweig G (1964) An SU_3 model for strong interaction symmetry and its breaking; Version 2. CERN-TH-412. Version 1 is CERN preprint 8182/TH.401
5. Standard model of elementary particles (2018). https://commons.wikimedia.org/wiki/File:Standard_Model_of_Elementary_Particles.svg. Accessed 3 Jan 2018
6. Fritzsch H, Gell-Mann M, Leutwyler H (1973) Advantages of the color octet gluon picture. Phys Lett B 47:365. https://doi.org/10.1016/0370-2693(73)90625-4
7. Gross DJ, Wilczek F (1973) Ultraviolet behavior of nonabelian gauge theories. Phys Rev Lett 30:1343. https://doi.org/10.1103/PhysRevLett.30.1343
8. Politzer HD (1973) Reliable perturbative results for strong interactions? Phys Rev Lett 30:1346. https://doi.org/10.1103/PhysRevLett.30.1346
9. 't Hooft G, Veltman MJG (1972) Regularization and renormalization of gauge fields. Nucl Phys B 44:189. https://doi.org/10.1016/0550-3213(72)90279-9
10. Glashow SL (1961) Partial symmetries of weak interactions. Nucl Phys 22:579. https://doi.org/10.1016/0029-5582(61)90469-2
11. Weinberg S (1967) A model of leptons. Phys Rev Lett 19:1264. https://doi.org/10.1103/PhysRevLett.19.1264
12. Salam A (1968) Weak and electromagnetic interactions. Conf Proc C 680519:367
13. Yang C-N, Mills RL (1954) Conservation of isotopic spin and isotopic gauge invariance. Phys Rev 96:191. https://doi.org/10.1103/PhysRev.96.191
14. Djouadi A (2008) The anatomy of electro-weak symmetry breaking. I: The Higgs boson in the standard model. Phys Rep 457:1. https://doi.org/10.1016/j.physrep.2007.10.004, arXiv:hep-ph/0503172
15. Gell-Mann M (1962) Symmetries of baryons and mesons. Phys. Rev. 125:1067. https://doi.org/10.1103/PhysRev.125.1067

16. Pauli W (1927) Zur Quantenmechanik des magnetischen Elektrons. Zeitschrift fur Physik 43:601. https://doi.org/10.1007/BF01397326
17. Englert F, Brout R (1964) Broken symmetry and the mass of gauge vector mesons. Phys Rev Lett 13:321. https://doi.org/10.1103/PhysRevLett.13.321
18. Higgs PW (1964) Broken Symmetries and the Masses of Gauge Bosons. Phys. Rev. Lett. 13:508–509. https://doi.org/10.1103/PhysRevLett.13.508
19. Guralnik GS, Hagen CR, Kibble TWB (1964) Global conservation laws and massless particles. Phys Rev Lett 13:585. https://doi.org/10.1103/PhysRevLett.13.585
20. t Hooft G (1971) Renormalizable Lagrangians for massive Yang-Mills fields. Nucl Phys B 35:167. https://doi.org/10.1016/0550-3213(71)90139-8
21. CDF Collaboration (1995) Observation of top quark production in $\bar{p}p$ collisions. Phys Rev Lett 74:2626. https://doi.org/10.1103/PhysRevLett.74.2626. arXiv:hep-ex/9503002
22. D0 Collaboration (1995) Search for high mass top quark production in $p\bar{p}$ collisions at $\sqrt{s} = 1.8$ TeV. Phys Rev Lett 74:2422. https://doi.org/10.1103/PhysRevLett.74.2422, arXiv:hep-ex/9411001
23. CMS Collaboration (2012) Observation of a new boson at a mass of 125 GeV with the CMS experiment at the LHC. Phys Lett B 716:30. https://doi.org/10.1016/j.physletb.2012.08.021. arXiv:1207.7235
24. ATLAS Collaboration (2012) Observation of a new particle in the search for the Standard Model Higgs boson with the ATLAS detector at the LHC. Phys Lett B 716:1. https://doi.org/10.1016/j.physletb.2012.08.020. arXiv:1207.7214
25. Goldstone J (1961) Field theories with superconductor solutions. Nuovo Cim 19:154. https://doi.org/10.1007/BF02812722
26. Goldstone J, Salam A, Weinberg S (1962) Broken symmetries. Phys Rev 127:965. https://doi.org/10.1103/PhysRev.127.965
27. Cabibbo N (1963) Unitary symmetry and leptonic decays. Phys Rev Lett 10:531. https://doi.org/10.1103/PhysRevLett.10.531
28. Kobayashi M, Maskawa T (1973) CP violation in the renormalizable theory of weak interaction. Prog Theor Phys 49:652. https://doi.org/10.1143/PTP.49.652
29. Glashow SL, Iliopoulos J, Maiani L (1970) Weak interactions with lepton-hadron symmetry. Phys Rev D 2:1285. https://doi.org/10.1103/PhysRevD.2.1285
30. Djouadi A (1997) Decays of the Higgs bosons. In: Quantum effects in the minimal supersymmetric standard model. Proceedings, international workshop, MSSM, Barcelona, Spain, p 197. arXiv:hep-ph/9712334
31. Romao JC, Andringa S (1999) Vector boson decays of the Higgs boson. Eur Phys J C 7:631. https://doi.org/10.1007/s100529801038. arXiv:hep-ph/9807536
32. LHC Higgs Cross Section Working Group (2018) https://twiki.cern.ch/twiki/bin/view/LHCPhysics/LHCHXSWG. Accessed 3-Jan-2018
33. ATLAS, CMS Collaboration (2016) Measurements of the Higgs boson production and decay rates and constraints on its couplings from a combined ATLAS and CMS analysis of the LHC pp collision data at $\sqrt{s} = 7$ and 8 TeV. JHEP 08:045. https://doi.org/10.1007/JHEP08(2016)045, arXiv:1606.02266
34. ATLAS, CMS Collaboration (2015) Combined Measurement of the Higgs boson mass in pp collisions at $\sqrt{s} = 7$ and 8 TeV with the ATLAS and CMS experiments. Phys Rev Lett 114:191803, https://doi.org/10.1103/PhysRevLett.114.191803, arXiv:1503.07589
35. CMS Collaboration (2018) Combined measurements of the Higgs boson's couplings at $\sqrt{s} =$ 13 TeV. CMS-PAS-HIG-17-031. https://cds.cern.ch/record/2308127
36. CMS Collaboration (2017) Measurements of properties of the Higgs boson decaying into the four-lepton final state in pp collisions at sqrt(s) = 13 TeV. JHEP 11:047. https://doi.org/10.1007/JHEP11(2017)047. arXiv:1706.09936
37. ATLAS Collaboration (2017) Measurement of the Higgs boson mass in the $H \rightarrow ZZ^* \rightarrow 4\ell$ and $H \rightarrow \gamma\gamma$ channels with $\sqrt{s}$=13TeV pp collisions using the ATLAS detector. ATLAS-CONF-2017-046. https://cds.cern.ch/record/2273853

38. CMS Collaboration (2018) Measurements of properties of the Higgs boson decaying to a W boson pair in pp collisions at $\sqrt{s} = 13$ TeV. CMS-PAS-HIG-16-042. https://cds.cern.ch/record/2308255
39. ATLAS Collaboration (2018) Measurement of gluon fusion and vector boson fusion Higgs boson production cross-sections in the $H \rightarrow WW^* \rightarrow e\nu\mu\nu$ decay channel in pp collisions at $\sqrt{s} = 13$ TeV with the ATLAS detector. ATLAS-CONF-2018-004. https://cds.cern.ch/record/2308392
40. CMS Collaboration (2018) Measurements of Higgs boson properties in the diphoton decay channel in proton-proton collisions at $\sqrt{s} = 13$ TeV. JHEP 11:185. https://doi.org/10.1007/JHEP11(2018)185. arXiv:1804.02716
41. ATLAS Collaboration (2017) Combined measurements of Higgs boson production and decay in the $H \rightarrow ZZ^{\rightarrow}4\ell$ and $H \rightarrow \gamma\gamma$ channels using $\sqrt{s} = 13$ TeV pp collision data collected with the ATLAS experiment. ATLAS-CONF-2017-047
42. CMS Collaboration (2018) Observation of the Higgs boson decay to a pair of τ leptons with the CMS detector. Phys Lett B 779:283. https://doi.org/10.1016/j.physletb.2018.02.004. arXiv:1708.00373
43. CMS Collaboration (2018) Inclusive search for a highly boosted Higgs boson decaying to a bottom quark-antiquark pair. Phys Rev Lett 120:071802. https://doi.org/10.1103/PhysRevLett.120.071802, arXiv:1709.05543
44. ATLAS Collaboration (2017) Evidence for the $H \rightarrow b\bar{b}$ decay with the ATLAS detector. JHEP 12:024. https://doi.org/10.1007/JHEP12(2017)024, arXiv:1708.03299
45. CMS Collaboration (2018) Evidence for the Higgs boson decay to a bottom quark-antiquark pair. Phys Lett B 780:501. https://doi.org/10.1016/j.physletb.2018.02.050, arXiv:1709.07497
46. ATLAS Collaboration (2016) Search for Higgs boson production via weak boson fusion and decaying to $b\bar{b}$ in association with a high-energy photon in the ATLAS detector. ATLAS-CONF-2016-063. https://cds.cern.ch/record/2206201
47. CMS Collaboration (2016) Search for the standard model Higgs boson produced through vector boson fusion and decaying to bb with proton-proton collisions at $\sqrt{s} = 13$ TeV. CMS-PAS-HIG-16-003. https://cds.cern.ch/record/2160154
48. ATLAS Collaboration (2017) Search for the dimuon decay of the Higgs boson in pp collisions at $\sqrt{s} = 13$ TeV with the ATLAS detector. Phys Rev Lett 119:051802. https://doi.org/10.1103/PhysRevLett.119.051802, arXiv:1705.04582
49. CMS Collaboration (2017) Search for the standard model Higgs boson decaying to two muons in pp collisions at $\sqrt{s} = 13$ TeV. CMS-PAS-HIG-17-019. https://cds.cern.ch/record/2292159
50. ATLAS Collaboration (2018) Evidence for the associated production of the Higgs boson and a top quark pair with the ATLAS detector. Phys Rev D97(7):072003. https://doi.org/10.1103/PhysRevD.97.072003. arXiv:1712.08891
51. CMS Collaboration (2018) Evidence for associated production of a Higgs boson with a top quark pair in final states with electrons, muons, and hadronically decaying τ leptons at $\sqrt{s} =$ 13 TeV. JHEP 08:066. https://doi.org/10.1007/JHEP08(2018)066. arXiv:1803.05485
52. ATLAS Collaboration (2018) Search for the standard model Higgs boson produced in association with top quarks and decaying into a $b\bar{b}$ pair in pp collisions at $\sqrt{s} = 13$ TeV with the ATLAS detector. Phys Rev D97(7):072016. https://doi.org/10.1103/PhysRevD.97.072016, arXiv:1712.08895
53. CMS Collaboration (2019) Search for $\mathrm{t\bar{t}H}$ production in the $\mathrm{H} \rightarrow \mathrm{b\bar{b}}$ decay channel with leptonic $\mathrm{t\bar{t}}$ decays in proton-proton collisions at $\sqrt{s} = 13$ TeV. JHEP 03:026. https://doi.org/10.1007/JHEP03(2019)026, arXiv:1804.03682
54. CMS Collaboration (2018) Observation of $\mathrm{t\bar{t}H}$ production. Phys Rev Lett 120(23):231801. https://doi.org/10.1103/PhysRevLett.120.231801, arXiv:1804.02610
55. Ng JN, Zakarauskas P (1984) A QCD parton calculation of conjoined production of Higgs bosons and heavy flavors in $p\bar{p}$ collision. Phys Rev D 29:876. https://doi.org/10.1103/PhysRevD.29.876

Chapter 3
Experimental Setup

The analysis uses proton-proton collisions to initiate the $t\bar{t}H$ process being searched for. The protons are accelerated and collided in the LHC and their collision and subsequent decay products are detected in the general purpose CMS detector, both of which are located at CERN. The CMS detector includes dedicated subsystems for measuring different particles and their properties, and combines their information in a propriety software system to obtain a full event description. Only events that may be of interest are fully reconstructed and saved for further analysis. This online event selection and recording is performed by the trigger and data acquisition systems, to both of which I made original contributions. My contributions include the development of a software package to interface the trigger with the data acquisition system, the estimation of trigger rates, the production of simulation samples for trigger testing and rate estimation, and contributing to the day-to-day operation and monitoring of the data acquisition system.

In this chapter, the experimental setup is described. It begins with the proton accelerator and collider and then moves on to the particle detector. The basic particle reconstruction techniques are described as well as the trigger system to select interesting events, and the data acquisition system to ensure those events are permanently recorded.

3.1 The Large Hadron Collider

The Large Hadron Collider (LHC) [1] is a dual-ring-superconducting-hadron accelerator and collider installed underground on the French-Swiss border near Geneva. It is located in the 26.7 km former LEP[1] tunnel that was originally constructed between 1984 and 1989. The tunnel consists of eight straight sections alternating with eight

[1]The Large Electron-Positron Collider (LEP) was a CERN e^+e^- accelerator and collider operating from 1989 to 2000. It serviced four complementary detectors: ALEPH, DELPHI, OPAL and L3.

D. Salerno, *The Higgs Boson Produced With Top Quarks in Fully Hadronic Signatures*, Springer Theses, https://doi.org/10.1007/978-3-030-31257-2_3

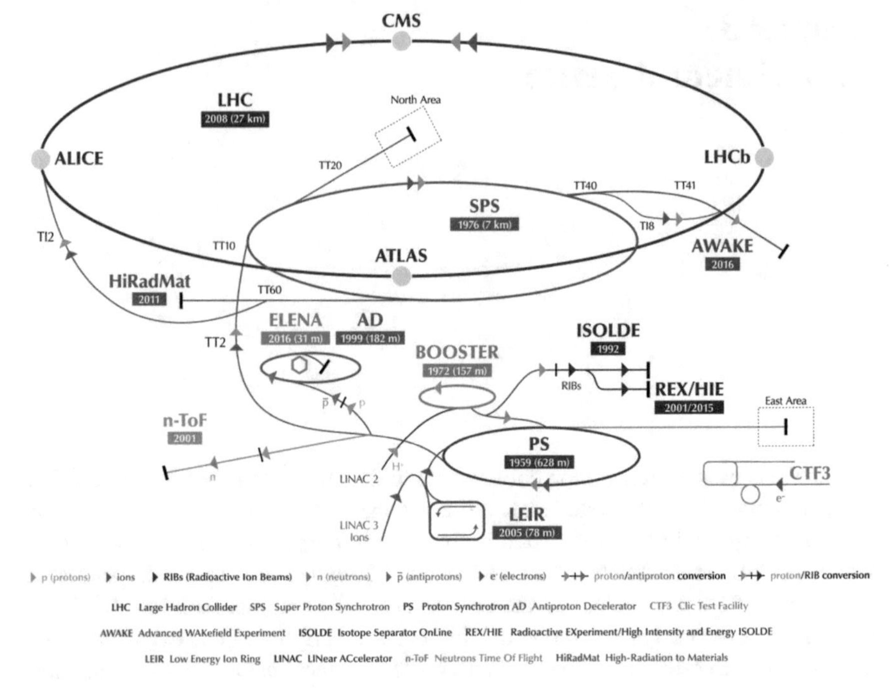

Fig. 3.1 Schematic overview of the CERN accelerator complex [2]. It shows the full accelerator chain, culminating in the LHC

arcs and lies below the surface at a depth of between 45 and 170 m, on an inclined plane with a 1.4% slope towards the Léman lake. It is the largest and most powerful particle accelerator ever built and is the gem of the CERN accelerator complex, shown in Fig. 3.1.

The LHC accelerates protons and heavy ions (X, predominantly lead nuclei) and can provide three types of collisions: p-p, X-X, and p-X. Since the topic of this thesis involves only proton-proton collisions, the following description of the LHC only considers protons. The protons for collision are taken from molecular hydrogen gas, separated and stripped of their electrons, before being injected in the linear accelerator LINAC2. The LINAC2 accelerates protons up to an energy of 50 MeV before injecting them into the proton synchrotron booster (PSB), which accretes them further up to an energy of 1.4 GeV. From the PSB, the protons are injected into the proton synchrotron (PS), which increases their energy to 25 GeV before injecting them into the super proton synchrotron (SPS). The SPS accumulates and accelerates the protons to an energy of 450 GeV, before injecting them in the LHC. The LHC accelerates the protons from 450 GeV up to a final design energy of 7 TeV, but as of now has only operated at up to 6.5 TeV per proton, resulting in centre-of-mass p-p collision energies of 13 TeV.

In each circular accelerator of the injector chain, the protons are accumulated in bunches equally spaced around the ring, with the number and intensity of the bunches

Table 3.1 LHC nominal proton beam parameters (Design) [1] and maximum achieved values (Actual) during 2016 data taking [3]

Parameter	Unit	Design	Actual
Injection energy	(GeV)	450	450
Collision energy	(GeV)	7000	6500
Instantaneous luminosity	($cm^{-2}s^{-1}$)	10^{34}	1.4×10^{34}
Number of bunches		2808	2208
Bunch spacing	(ns)	24.95	25
Intensity per bunch	(p/b)	1.15×10^{11}	1.15×10^{11}
Beam current	(A)	0.58	0.46
Transverse emittance (RMS, normalised)	(μm)	3.5	2.0
Longitudinal emittance (total)	(eVs)	2.5	0.6
Bunch length (4σ)	(ns)	1.0	1.1
Energy spread (4σ)	(10^{-3})	0.45	–

increasing at each stage. When all bunches have been injected into the LHC, the LHC then begins to accelerate the protons to their final collision energy. The time period for which the LHC has all bunches circulating in a beam is called a *fill*. The nominal design parameters for the LHC beam are listed in Table 3.1, along with the actual operating performance in 2016.

The proton beams are made to collide at four points around the LHC ring, where four main detectors are located:

- **ALICE**: a dedicated heavy ion detector [4].
- **ATLAS**: a general purpose high luminosity detector [5].
- **CMS**: a general purpose high luminosity detector, see Sect. 3.2.
- **LHCb**: a low luminosity detector dedicated to b-physics [6].

In addition, there is a low luminosity detector TOTEM [7], which aims to measure the total proton-proton cross section based on the collisions at the CMS interaction point.

Although the proton bunches intersect at every bunch crossing, relatively few of the protons actually collide. In most cases, protons will merely "skim" another proton in what is known as an *elastic collision*, in which the proton structure is unaltered. The interesting physics occurs when a proton undergoes a head-on or *inelastic collision* that permanently alters the proton as its quark or gluon constituents interact with the constituents of another proton. The probability for these interesting collisions to occur is related to the *cross section* of the given process. With this in mind, the number of interactions per second (rate) generated in the LHC collisions is given by:

$$R_{\text{proc}} = L\sigma_{\text{proc}}, \tag{3.1}$$

where σ_{proc} is the cross section for the process under study and L is the LHC instantaneous luminosity. The instantaneous luminosity depends on the beam parameters and, for a Gaussian beam distribution, can be written as:

$$L = \frac{N_b^2 n_b f_{\text{rev}} \gamma_r}{4\pi \epsilon_n \beta^*} F, \tag{3.2}$$

where N_b is the number of protons per bunch, n_b is the number of bunches per beam, f_{rev} is the frequency of revolution, γ_r is the relativistic gamma factor, ϵ_n is the normalised transverse beam emittance, β^* is the beta function at the interaction point (IP), and F is a reduction factor due to the crossing angle at the IP, defined as:

$$F = 2\left(1 + \left(\frac{\theta_c \sigma_z}{2\sigma^*}\right)^2\right)^{-1/2}, \tag{3.3}$$

where θ_c is the crossing angle at the IP, σ_z is the RMS bunch length, and σ^* is the transverse RMS beam size at the IP. Equation (3.3) assumes both beams are round and have equal beam parameters. In addition to the high energies required to initiate rare physics processes, a high beam intensity is essential to ensure many such rare events are produced. The nominal instantaneous luminosity is typically achieved at the start of a fill, but as time progresses, the intensity of protons in each bunch decreases, not only from the p-p interactions at each collision point, but also through interactions with the beam gas or accelerator material, and losses from protons escaping the field of the LHC bending magnets. Typically a fill is maintained for several (up to around 30) hours before the instantaneous luminosity decreases to such a low rate that it becomes beneficial to dump the beam and re-fill, accepting the two-hour downtime between fills. Occasionally, the LHC loses control of the beam or an emergency arises that requires the beam to be dumped mid-fill.

The total proton-proton cross section at a centre-of-mass energy of $\sqrt{s} = 13$ TeV is expected to be approximately 70 mb, which means that around 20 p-p collisions will occur at each bunch crossing at the design luminosity, in addition to any event of interest. One of the main challenges facing a high-luminosity experiment such as CMS is therefore to disentangle the particles coming from the event of interest from those originating from the more common inelastic p-p collisions, referred to as pileup.

3.2 The CMS Detector

The Compact Muon Solenoid (CMS) detector is a multi-purpose apparatus operating at the LHC. It is located at the opposite end of the LHC ring from the main CERN site, near the French village of Cessy, at about 100 meters underground. It is housed in the experimental cavern which is separated from the neighbouring service cavern

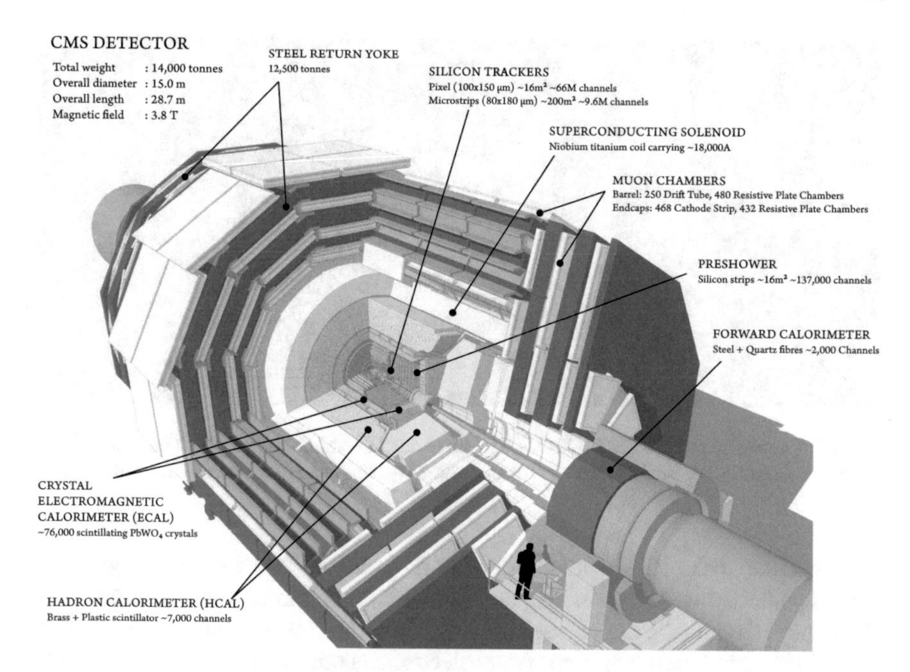

Fig. 3.2 3-dimensional schematic of the CMS detector with annotations [10]

by a thick concrete wall, allowing radiation-free access to the service cavern during LHC operation. CMS is constructed in semi-circular slices that surround the LHC beam pipe, which runs through its centre, creating a cylindrical form of 21.6 m length, 15.0 m diameter, and 14 000 t weight.

CMS uses a coordinate system which has its origin centred at the nominal LHC collision point (interaction point), the y-axis pointing vertically upward, the x-axis pointing toward the centre of the LHC ring, and thus the z-axis pointing in the anticlockwise direction of the beam. A mixture of cartesian, cylindrical and spherical coordinates are used, with each coordinate adopting a unique definition. The azimuthal angle ϕ is measured in the x-y plane relative to the x-axis, and r is the radial coordinate in this plane. The polar angle θ is measured from the z-axis, leading to the definition of pseudorapidity $\eta = -\ln[\tan(\theta/2)]$. The x and y components of momentum and energy are used to determine these quantities transverse to the beam axis, p_{T} and E_{T}, respectively.

A schematic diagram of the CMS detector is shown in Fig. 3.2, while a photograph in its open position is shown in Fig. 3.3. The central feature of CMS is a large superconducting solenoid of 12.5 m length and 6.3 m internal diameter, providing a uniform magnetic field of 3.8 T in its centre. The 220 t cold mass operates at a temperature of approximately 4.6 K and includes a 4-layer winding of reinforced NbTi conductor housed in an aluminium alloy, holding a stored energy of 2.6 GJ at full current. The magnetic field generated by the solenoid has a bending power of 12 Tm and is returned through a 12 500 t iron yoke, composed of 5 central slices

Fig. 3.3 Photograph of the CMS detector with an endcap open [22-Mar-2017]. The central barrel can be seen on the left, the LHC beam pipe in the centre and the negative-z end cap on the right

and 6 end cap disks. The bending power of the solenoid is key to providing strong separation between charged and neutral particles as well as accurate measurements of the momentum of charged particles. Within the solenoid reside a silicon pixel and strip tracker, a lead tungstate crystal electromagnetic calorimeter (ECAL), and a brass and scintillator hadron calorimeter (HCAL), each composed of a barrel and two endcap sections. Forward calorimeters extend the pseudorapidity coverage provided by the barrel and end detectors. Outside the solenoid, gas-ionisation muon detectors are embedded in the return yoke. Further details of each subdetector are provided below, while full details of the CMS detector and its components can be found in Ref. [8]. Details of the reconstruction algorithms of each subdetector are also included below and are primarily taken from Ref. [9].

3.2.1 Silicon Pixel and Strip Trackers

The inner tracking system of CMS is composed of two separate silicon-based detectors which together provide an accurate and efficient measurement of the charged particles produced in the collisions, as well as a precise reconstruction of particle origins, known as vertices. The innermost component is the pixel detector, which in 2016 consisted of three barrel layers and two endcaps, each with two disks. Around the pixel detector lies the silicon strip tracker, which has 10 barrel layers and two endcaps with 3 small and 9 large disks each. The total dimensions of the tracker are 5.8 m length and 2.5 m radius, with about 200 m^2 of active silicon, providing an acceptance of up to $|\eta| = 2.5$.

The requirements of the tracking system are very demanding, requiring a reliable and precise reconstruction of charged particle trajectories, in a high density region of activity. At the design luminosity of the LHC, on average around 1000 particles

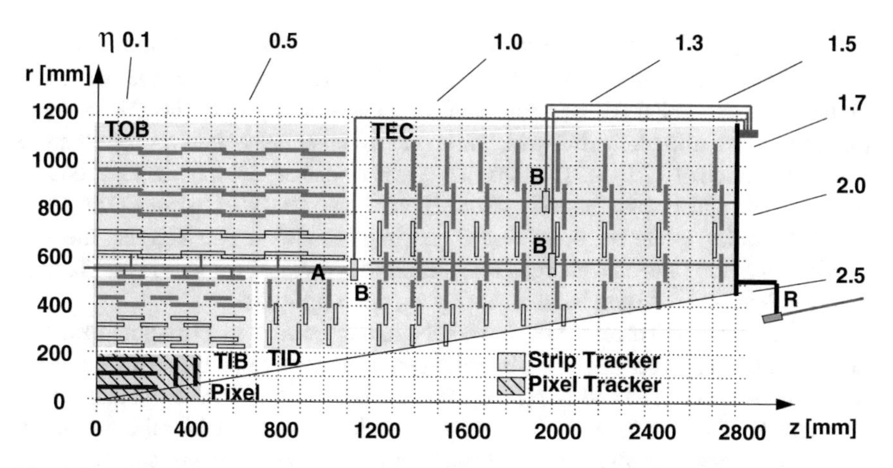

Fig. 3.4 Schematic cross section of one quadrant of the CMS tracker [11]. Single silicon strip module positions are indicated as solid lines, double strip modules as open lines, and pixel modules as solid lines. Also shown are the paths of the laser rays (R), the beam splitters (B), and the alignment tubes (A) of the Laser Alignment System (not discussed)

hit the tracker every 25 ns bunch crossing. The total particle rate of 40 GHz implies a hit rate density of 1 MHz/mm^2 at a radius of 4 cm, 60 kHz/mm^2 at a radius of 22 cm and 3 kHz/mm^2 at a radius of 115 cm. In order to ensure the occupancy[2] remains below 1%, high-density pixelated detectors are required at radii below 10 cm, while micro-strip detectors can be used at radii between 20 and 55 cm, and larger strips can be used in the outer region of the tracker. A schematic overview of the inner tracking system, which is symmetric with respect to the z-axis and the x-y plane, is shown in Fig. 3.4.

The pixel detector measures particles closest to the interaction point and is instrumental in the reconstruction of the primary interaction vertex. It consists of three cylindrical layers at radii of 4.4, 7.3 and 10.2 cm and two disks on each side, at 34.5 and 46.5 cm from the interaction point. There are approximately 66 million pixels with dimensions of $100 \times 150\,\mu\text{m}^2$, housed on a total of 1440 sensor modules, with an active area of around 1 m^2. The pixel dimensions ensure a similar track resolution in both r-ϕ and z directions and deliver three high-precision points on each charged particle track with a pseudorapidity range of $|\eta| < 2.5$. In the barrel layers, the electron drift to the collecting pixel implant is perpendicular to the magnetic field, thus a Lorentz drift leads to charge spreading across neighbouring pixels. The readout of an analogue pulse height then allows a charge interpolation to be made, which results in a spatial resolution considerably smaller than the pixel dimensions, of around 15–20 μm.

The silicon strip tracker covers the region between a radius of 20 and 116 cm and is composed of three different subsystems. The tracker inner barrel (TIB) and disks (TID) consists of four barrel layers of 1.4 m length and radii of up to 50 cm and

[2] Occupancy refers to the proportion of sensors that are hit per bunch crossing

three disks at each end, from 75 to 105 cm in the z direction. The TIB and TID use 320 μm thick silicon micro-strip sensors with their strips parallel to the beam axis in the barrel and radial on the disks. The two innermost layers of the TIB host two modules with a strip pitch of 80 μm, while the two outer layers host a single module with a strip pitch of 120 μm. The three TIDs at each endcap are identical and consist of three rings which span the radius from 20 to 50 cm. The two innermost rings host double modules, while the outer ring hosts single modules. The TIB and the TID provide up to four hits on a charged track, each with a spatial resolution of 23 μm in the inner layers and 35 μm in the outer layers, up to $|\eta| < 2.5$.

The tracker outer barrel (TOB) covers the radius from 50 to 116 cm and extends to $z = \pm 118$ cm. It consists of six barrel layers of 500 μm thick micro-strip sensors with strip pitches of 183 and 122 μm on the first four layers and the two outer layers, respectively. The two innermost layers host double modules, while the four outer layers have single modules. The TOB provides up to six r-ϕ hits on a charged track, with a single point resolution of 53 μm in the inner four layers and 35 μm in the outer layers.

Beyond the z range of the TOB, the tracker endcaps (TECs) provide coverage for $124 < |z| < 282$ cm and $22.5 < |r| < 113.5$ cm. Each TEC consists of nine disks composed of four to seven rings of silicon micro-strip detectors, with a thickness of 320 μm on the four inner rings and 500 μm on the outer rings. The two innermost rings and the fifth ring have double modules, while the other rings (3, 4, 6, and 7) have single modules. The strips are placed radially with an average pitch of 97 to 184 μm, providing up to nine hits per charged track with $|\eta| < 2.5$.

The double modules mentioned above are placed with a stereo angle of 100 mrad to provide a measurement of the second coordinate (z in the barrel or r on the disks). The single point resolution of this measurement is 230 μm in the TIB and 530 μm in the TOB, and varies with the strip pitch in the TID and TEC. The layout of the tracker provides at least nine hits in the silicon strip tracker up to a pseudorapidity range of $|\eta| < 2.5$ with at least four of them being two-dimensional measurements. In total, there are about 9.3 million strips with an active silicon area of 198 m^2.

The several layers of active tracker material, together with the passive material such as support, cables and cooling, give rise to particle interactions before reaching the calorimeters. The amount of material[3] as a function of η, broken down by component, is shown in Fig. 3.5. At the maximum thickness ($|\eta| \approx 1.5$), there is about an 85% probability that a photon will convert to an e^+e^- pair or an electron will radiate a photon by interacting with this material. At the same trajectory, there is roughly a 20% probability that a hadron will interact with the material. The large number of secondary particles produced in these interactions with the tracker material, pose a challenge to the particle reconstruction, which is overcome by exploring the full granularity and redundancy of the tracker measurements.

[3]The amount of material is expressed in units of radiation length X_0 and interaction length λ_t. X_0 is characterised by electromagnetic interactions and is the mean distance over which an electron loses all but $1/e$ of its energy through bremsstrahlung. λ_t is characterised by nuclear interactions and is the mean distance required to reduced the number of charged particles by a factor of $1/e$.

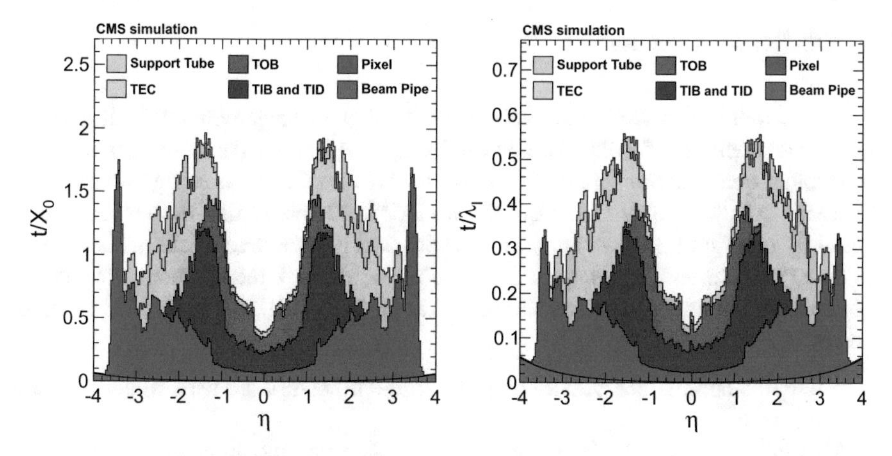

Fig. 3.5 Total thickness t of the inner tracker material expressed in units of radiation length X_0 (left) and interaction length λ_t (right), as a function of η and broken down by component [12]

Tracker Readout

The pixel detector read-out and control system is formed by a three-part chain: a data read-out link from the modules to the pixel front end driver (pxFED), a fast control link from the pixel front end controller (pFEC) to the modules, and a slow control link to configure the readout electronics. The sensor signals are read out by read-out chips (ROCs) which are custom ASICs[4] bump bonded to 52×80 pixels. Several ROCs are controlled and read-out by a token bit manager (TBM) which sends an analogue signal to a pxFED, which then digitises and formats it before sending it to the data acquisition system (DAQ), described in Sect. 3.4. At the same time, a pFEC sends the 40 MHz clock and fast control signals, such as trigger and reset signals, to each TBM over a digital link. The pFECs and pxFEDs are located in the service cavern and connected to the TBMs by 40 MHz optical links.

The readout system for the strip tracker is slightly simpler than for the pixel detector: a data read-out link from the silicon sensors to the front end driver (FED) and a control link from the front end controller (FEC) to the sensors. The sensor signals are amplified, shaped and stored by a custom ASIC, which, upon a positive trigger decision, sends the analogue signals to a FED, which then digitises them and at 40 MHz sends them to the DAQ. The clock, trigger and control signal are transmitted by optical links from the FECs to the custom ASICs. The FEDs and FECs are located in the service cavern at a distance of about 100 m from the tracker and are connected with analogue and digital optical links, at 40 MHz and 40 Mb/s, respectively.

[4]An application-specific integrated circuit (ASIC) is an integrated circuit designed for a particular use.

Track Reconstruction

The basic motivation driving the reconstruction of charged-particle tracks, is to measure the momentum of isolated muons, to identify hadronic τ decays, and to identify jets from b-quark hadronisation. While the track reconstruction is straightforward for energetic particles with well-measured tracks, CMS can also utilise the tracker to successfully measure lower energy particles with poorly-measured tracks, as described in Sect. 4.3.1. In both cases, the starting point for all track reconstruction is a combinatorial track finder based on Kalman filtering (KF) [13]. It is performed in three stages:

- A few hits compatible with a charged-particle trajectory are used to generate an initial seed.
- All hits from all tracker layers along this charged-particle trajectory are gathered in what is called trajectory building (or pattern recognition).
- A final fit is performed to determine the charged-particle properties: origin, transverse momentum, and direction.

Stringent track quality criteria are applied when this is the final reconstruction method: the seed must include two hits in consecutive layers of the pixel detector; there must be at least eight hits in total, with each contributing less than 30% of the overall track goodness-of-fit χ^2, and with at most one missing hit along the way; and all tracks must originate from within a few mm of the beam axis and have p_T greater than 0.9 GeV.

The performance of this track finder in terms of reconstruction efficiency of charged tracks and misreconstruction rate of wrong tracks, is about 70–80% efficiency for charged pions with $p_T > 1$ GeV and 99% efficiency for isolated muons, and a few percent misreconstruction rate for pions. The difference between pion and muon efficiency is primarily due to nuclear interactions with the tracker material, which can be inferred from Fig. 3.5 (right) and ranges from 10 to 30%. The tracking efficiency is reduced for high-p_T particles ($>$ 10 GeV), which are often found in collimated jets and thus the presence of overlapping particles makes it difficult to identify the correct tracks. As will be discussed in Sect. 4.3.1, significant improvements on this performance are ultimately achieved by CMS.

3.2.2 Electromagnetic Calorimeter

The ECAL is a hermetic and homogenous calorimeter made from 61 200 lead tungstate ($PbWO_4$) crystals in the central barrel ($|\eta| < 1.479$) and 7 324 crystals in each endcap ($1.479 < |\eta| < 3.0$). The ECAL barrel has an inner radius of 129 cm and is composed of 36 identical wedge-shaped "supermodules" covering half the barrel length. The endcaps are placed at $z = \pm 314$ cm and are formed by two semi-circular aluminium plates containing 5×5 crystal units, "supercrystals".

The crystals induce an electromagnetic shower of light with a Molière radius[5] of 2.2 cm. The crystal length is 23 cm in the barrel and 22 cm in the endcaps, corresponding to radiation lengths of 25.8 and 24.7, respectively, which is sufficient to contain over 98% of the energy of electrons and photons up to 1 TeV. This length of crystals also corresponds to about one interaction length, which implies that around two thirds of hadrons will start showering in the ECAL before entering the HCAL. The scintillation light produced in the shower results in around $30\,\gamma/\text{MeV}$ and is measured by avalanche photodiodes in the barrel and vacuum phototriodes in the endcap, both with intrinsic gain and able to operate in a magnetic field.

The transverse size of the crystals is $2.2 \times 2.2\,\text{cm}^2$ in the barrel and and $2.9 \times 2.9\,\text{cm}^2$ in the endcaps. This fine transverse granularity is similar to the Molière radius, thus allowing hadron and photon energy deposits as close as 5 cm to be resolved. The intrinsic energy resolution of the ECAL barrel is measured to be [14]:

$$\frac{\sigma}{E} = \frac{2.8\%}{\sqrt{E}} \oplus \frac{12\%}{E} \oplus 0.3\% \tag{3.4}$$

where E is expressed in GeV, the first term on the right side is the stochastic term, the second term is the noise and the last term is a constant. The small stochastic term ensures that the photon energy resolution is excellent in the typical range of photons in jets (1–50 GeV).

The electronics noise in the ECAL is measured to be around 40 and 150 MeV in the barrel and endcap, respectively, and is suppressed offline by requiring each crystal to have an energy in excess of twice this noise term. Another source of spurious signals is from particles that directly ionise the photodiodes used to collect the scintillation light, which can be rejected by requiring compatible energy deposits in neighbouring crystals and timing within 2 ns of the beam crossing for high energy (>1 GeV) deposits.

A finer grained detector, called the *preshower*, is installed in front of each ECAL endcap, made of two layers of lead radiator followed by silicon strip sensors. Initially it was intended to identify photons from π^0 decays to discriminate them from prompt[6] photons, however the large number of neutral pions produced by hadronic interactions with the tracker material substantially reduce the preshower's identification capability. In current operations, the energy deposited in the preshower is simply added to that of the closest associated ECAL crystals.

[5]The Molière radius is defined as the radius of a cylinder containing 90% of the energy deposition of the electromagnetic shower.

[6]Prompt particles refer to those produce in the primary pp interaction and not the subsequent decays.

ECAL Readout

The ECAL electronics is divided into two subsystems: the *front-end* electronics, composed of radiation-resistant circuits positioned immediately behind the crystals, and the *back-end* electronics, located in the service cavern. The two systems are connected by 90 m long high-speed optical links with a bandwidth of 800 Mb/s.

The front-end electronics are formed by grouping 5×5 crystals into blocks, called trigger towers in the barrel and supercrystals in the endcaps. Each block contains electronics connected to the photodiodes/phototriodes in groups of five crystals, which amplify and digitise the signals at 40 MHz, buffer the data until a trigger decision is received, and finally transmit the data to the back-end electronics. In addition, each block creates trigger primitives from the digitised data and transmits them, via the back end, at 40 MHz to the Level-1 trigger, described in Sect. 3.3.1.

The back-end electronics connect the ECAL to both the trigger and the DAQ systems. For the trigger, at each bunch crossing, the trigger primitives generated in the front-end electronics are finalised and synchronised in a trigger concentration card, before being sent to the regional calorimeter trigger. For the DAQ, the data from the front end is read out and reduced by the data concentration card, based on the selective readout flags that determine which sectors are to be read out and at which level of suppression.

ECAL Reconstruction

The energy deposited in the ECAL crystals is generally spread out over a few neighbouring crystals, such that the total energy is measured in several crystals. A specific clustering algorithm was developed by CMS with four specific purposes:

- to measure the energy and direction of stable neutral particles, i.e. photons and neutral hadrons;
- to separate these neutral particles from charged hadron energy deposits;
- to identify and reconstruct electrons and accompanying bremsstrahlung photons; and
- to supplement the energy measurement of charged hadrons which cannot be accurately measured by the tracker, e.g. for low-quality and high-p_{T} tracks.

The clustering algorithm is the same for the ECAL, preshower and HCAL, and is performed separately in the barrel and endcaps of each subdetector. In the following a cell refers to an ECAL barrel tower, an ECAL endcap supercrystal, a preshower silicon strip or an HCAL tower.

The algorithm begins by identifying *cluster seeds* as local maxima of calorimeter-cell energy with respect to the four or eight surrounding cells, provided they have an energy above a given seed threshold. Then cells are aggregated to form *topological clusters*, by adding cells with an energy above a given cell threshold and at least a corner in common with a cell already in the cluster. In the ECAL endcaps, seeds are additionally required to satisfy a threshold on E_{T}, because of the increased noise at high θ.

Finally, an expectation-maximisation algorithm algorithm based on a Gaussian-mixture model is used to reconstruct the resulting clusters within a topological cluster. The model postulates that the energy deposits in the M cells of the topological cluster arise from N Gaussian energy deposits, where N in the number of seeds in the topological cluster. The model returns two parameters: the amplitude A_i and the coordinates in the η–ϕ plane of the mean of each Gaussian μ_i, while the width of the Gaussian is fixed to different values depending on the calorimeter. The amplitude is constructed such that $\sum A_i = \sum E_j$, where E_j is the energy measured in cell j of the topological cluster. After convergence, the position and amplitude of the Gaussian functions are taken as the position and energy of the clusters.

The clustering algorithm is illustrated in Fig. 3.6, which shows an event display of five particles in a jet. In Fig. 3.6c, two cluster seeds (dark grey) are present in the HCAL within one topological cluster of nine cells. Following the fit, the two seeds result in two HCAL clusters, the final positions of which are indicated by the round points. These reconstructed positions are very close to the two charged-pion track extrapolations from the tracker. Similarly, the ECAL topological cluster of Fig. 3.6b stemming from the π^0 is split in two clusters corresponding to the two photons from its decay.

ECAL Energy Calibration

The energy of photons and neutral hadrons can only be obtained by measurements in the calorimeters, as they do not leave any trace in the tracker. While this is relatively straightforward for isolated neutral particles, a complication arises when neutral particles overlap with charged particles. In this case, the energy deposits of the neutral particle can only be detected as a calorimeter energy excess over the sum of charged particle momenta obtained from the tracker. An accurate calibration of the calorimeter response to neutral particles, and also charged particles, is crucial to maximising the probability of identifying neutral particles and determining their energy, while minimising the rate of misreconstructed energy excesses. The calibration is also important to correct for threshold effects, in which the energy deposit in cells is ignored unless it is above a certain threshold.

The ECAL calibration was initially made prior to the first LHC collisions, and then refined with collision data at centre-of-mass energies of 7, 8 and 13 TeV. The calibration necessitates corrections to the measured calorimeter energies, which are derived with simulated photons. In the ECAL barrel, the following correction is applied:

$$E_{\text{calib}} = f(E, \eta) E_{\text{ECAL}} = g(E) h(\eta) E_{\text{ECAL}}, \tag{3.5}$$

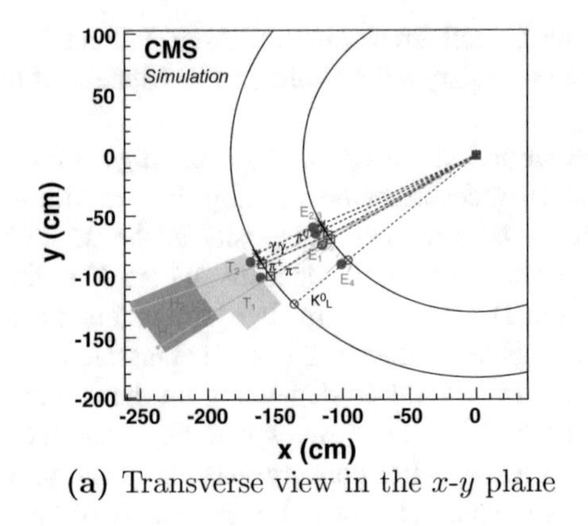

(a) Transverse view in the x-y plane

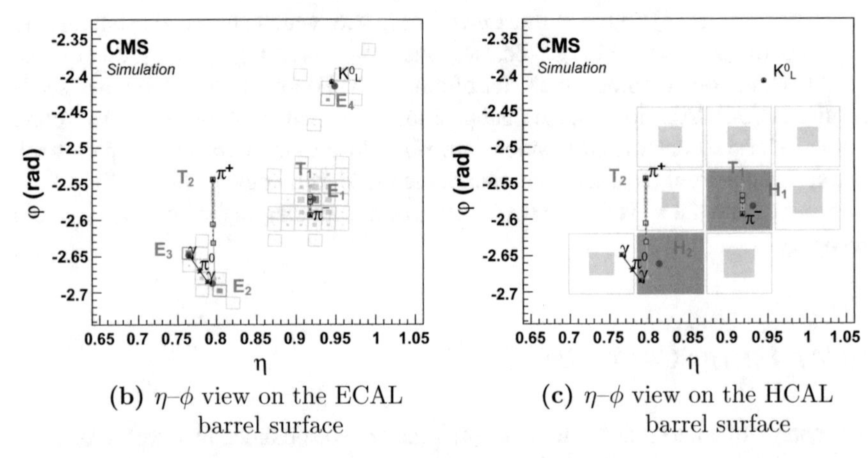

(b) η–ϕ view on the ECAL barrel surface

(c) η–ϕ view on the HCAL barrel surface

Fig. 3.6 Event display of an illustrative jet made of just five particles [9]. In (**a**), the ECAL and HCAL surfaces are represented as circles centred around the interaction point. The K_L^0, the π^-, and the two photons from the π^0 decay are detected as four well-separated ECAL clusters denoted $E_{1,2,3,4}$. The π^+ does not create a cluster in the ECAL. The two charged pions are reconstructed as charged-particle tracks $T_{1,2}$, appearing as vertical solid lines in the η–ϕ views and arcs in the x-y view. These tracks point towards two HCAL clusters $H_{1,2}$. In (**b**) and (**c**), the ECAL and HCAL cells are represented by squares, with an inner shaded area proportional to the logarithm of the cell energy. Cells with an energy larger than those of the neighbouring cells are shown in dark grey. In all three views, the fitted cluster positions are represented by round points, the simulated particles by dashed lines, and the positions of their impacts on the calorimeter surfaces by open square markers

where E and η are the energy and pseudorapidity of the cluster. The function $f(E, \eta) = g(E)h(\eta)$ is fitted to a two-dimensional distributional of the average ratio of the true photon energy E_{true} to the cluster energy, $\langle E_{\text{true}}/E \rangle$. The correction is close to one at high energy, where threshold effects effectively vanish, while it can be up to 1.2 (+20%) at low energy.

In the ECAL endcaps, the measured calorimeter energy includes the energy deposited in the two preshower layers, E_{PS1} and E_{PS2}. The calibrated energy is therefore expressed as:

$$E_{\mathrm{calib}} = \alpha(E, \eta)E_{\mathrm{ECAL}} + \beta(E, \eta)E_{\mathrm{PS1}} + \gamma(E, \eta)E_{\mathrm{PS2}}, \tag{3.6}$$

where E and η are now the energy and pseudorapidity of the generated photon. The calibration parameters α, β, and γ are chosen in each (E, η) bin to minimise a χ^2 on the difference between E_{calib} and E. In the region beyond the preshower acceptance or when no energy is measured in the preshower, the correction is applied as in Eq. (3.5). In the fiducial region of the preshower, the fitted parameters correct the ECAL energy by up to +40% for the smallest photon energies and by +5% at the largest photon energies, implying that an energetic photon loses an average of 5% of its energy in the preshower material. In all ECAL regions and for all energies, the calibrated energy agrees with the true photon energy to within ±1% on average.

3.2.3 *Hadron Calorimeter*

The HCAL is a hermetic sampling calorimeter made from several layers of brass absorber and plastic scintillator tiles, surrounding the ECAL. It has a barrel with an acceptance of $|\eta| < 1.4$ and two endcap disks covering $1.3 < |\eta| < 3.0$, and is complemented by the hadron outer (HO) sitting outside the solenoid. The HO is a single layer of 10 mm thick scintillators, corresponding to 1.4 interaction lengths at normal incidence, covering the region $|\eta| < 1.26$, and serves as a "tail catcher" of hadronic showers leaking through the calorimeters. In the very central region ($|\eta| < 0.25$), a 20 cm layer of steel increases the thickness of the HO to a total of 3 interaction lengths. The total thickness of the ECAL+HCAL calorimeter system is about 12 and 10 interaction lengths in the barrel and endcaps, respectively.

The HCAL scintillating tiles are connected to multi-channel hybrid photodiodes, with a gain of around 2000, by embedded wavelength-shifting fibres spliced to clear fibres outside the scintillator. The scintillating tiles have a thickness of 3.7 mm and are inserted in the overlapping brass plates, except for the first layer in the barrel which sits in front of the brass and is 9 mm thick. The absorber-scintillator layers are grouped into segments called *towers*. The HCAL barrel is formed of two half barrels, each composed of 18 identical wedges covering half the pseudorapidity region $|\eta| < 1.4$, with each wedge consisting of 4 rows of 18 towers, with a segmentation of $\Delta\eta \times \Delta\phi = 0.087 \times 0.087$. Each HCAL endcap disk is formed of semi-circular brass plates, in between which reside 17 layers of scintillating tiles [15]. The endcap segmentation is 5° in ϕ and 0.087 in η for the five outmost towers (smallest η) and 10° in ϕ and from 0.09 to 0.35 in η for the eight/nine innermost towers.

The energy resolution of the HCAL has been measured to be [16]:

$$\frac{\sigma}{E} = \frac{110\%}{\sqrt{E}} \oplus 9\% \tag{3.7}$$

where E is expressed in GeV. The electronics noise in the HCAL is measured to be around 200 MeV per tower. Additional, high-amplitude, coherent noise occurs rarely in an entire row or wedge of towers in the barrel, which can be easily rejected offline.

An additional calorimeter, the hadron forward (HF), is situated beyond the muon chambers at $z = \pm 11$ m and extends the pseudorapidity coverage on both sides up to $|\eta| \simeq 5$. It consists of steel absorbers with embedded radiation-hard quartz fibres running parallel to the beam, which alternate between full length fibres (about 165 cm or 10 interaction lengths) and shorter fibres starting 22 cm from the front face and exiting at the back. The long and short fibres are grouped into towers with a segmentation of $\Delta\eta \times \Delta\phi = 0.175 \times 0.175$ over most of the acceptance, each of which is connected to two photomultipliers.[7] The HF towers are used to estimate the electromagnetic and hadronic components of the shower, by acknowledging that most of the electromagnetic energy deposit is concentrated in the first 22 cm of the absorber. Then, if L and S denote the energy measured in the long and short fibres of the tower, respectively, the electromagnetic energy component can be approximated by $L - S$, while the hadron component is $2S$. Spurious signals in the HF, caused by high energy particles directly hitting the photomultiplier windows, can be rejected by requiring certain compatibility between L and S energy deposits, timing restrictions, and comparing neighbouring towers.

HCAL Readout

The light from the scintillation material is collected in photodiodes in the HCAL and HO and photomultipliers in the HF, which convert it to an electrical signal before passing it to an analogue-to-digital converter (ADC). The ADC transmits the 32-bit digital output at every bunch crossing (40 MHz) to a gigabit optical link chip, which then sends it to the service cavern via 1.6 Gb/s optical fibres.

In the service cavern, the incoming data is deserialised and processed by the HCAL trigger readout board (HTR). The HTR constructs the trigger primitives and sends them to the regional calorimeter trigger. It also buffers the full readout data waiting for a trigger accept signal and then transmits it to the DAQ system via the data concentration card.

[7] Conventional photomultiplier tubes can be used in this forward region where the magnetic field is much weaker than in the central detector.

HCAL Reconstruction

The energy form the HCAL towers is clustered in the same way as for all the calorimeters, described in Sect. 3.2.2. In the HF however, no clustering is performed and the electromagnetic and hadronic components of each tower are used directly to construct an HF EM cluster and HF HAD cluster, respectively.

HCAL Energy Calibration

The energy of hadrons is generally deposited in both the ECAL and the HCAL. The ECAL energy calibration described in Sect. 3.2.2 gives the correct energy for photons, but not for hadrons, which have a substantially different energy profile. The HCAL was initially calibrated with a pion test beam with no ECAL interaction, however the actual HCAL response depends on the fraction of energy deposited in the ECAL, which varies non-linearly with energy. Therefore, a recalibration of the energies from the ECAL and HCAL clusters is necessary for an accurate estimate of the true hadron energy.

The calibrated calorimetric energy of a hadron is calculated as:

$$E_{\text{calib}} = a + b(E) f(\eta) E_{\text{ECAL}} + c(E) g(\eta) E_{\text{HCAL}}, \tag{3.8}$$

where E and η are the true energy and pseudorapidity of the hadron. The constant coefficient a is expressed in GeV and accounts for threshold effects of the clustering algorithm. The coefficients a, b, and c, and the functions f and g are determined with simulated K^0_L events, by iteratively minimising a χ^2 in bins of E. The determination is made separately in the barrel and endcaps, and separately for hadrons leaving energy in both the ECAL and HCAL and those depositing energy solely in the HCAL. Hadrons leaving energy only in the ECAL are not calibrated, as such clusters would be considered photons or electrons.

The constant a is chosen to minimise the dependence of b and c on E, for energies above 10 GeV. Its value is set to 1.2 GeV for hadrons showering only in the HCAL, and 3.5 GeV for those showering in both the ECAL and HCAL. The calculated values of the coefficients b and c in each energy bin of the barrel region is shown in Fig. 3.7a.

The calibrated energy is used to calculate the *calibrated response*,[8] while the cluster energy is used to calculate the *raw response*. These quantities are displayed in Fig. 3.7b, along with the associated energy resolution. The effect of the calibration is to successfully bring the response close to zero for all energies, and substantially improve the resolution for low energies. The improved energy resolution below 10 GeV is a result of the coefficients b and c going to to zero at low energy and is explained as follows. Hadrons with true energy below 10 GeV typically do not

---

[8] The energy response is defined as the mean relative difference between the measured energy and the true energy.

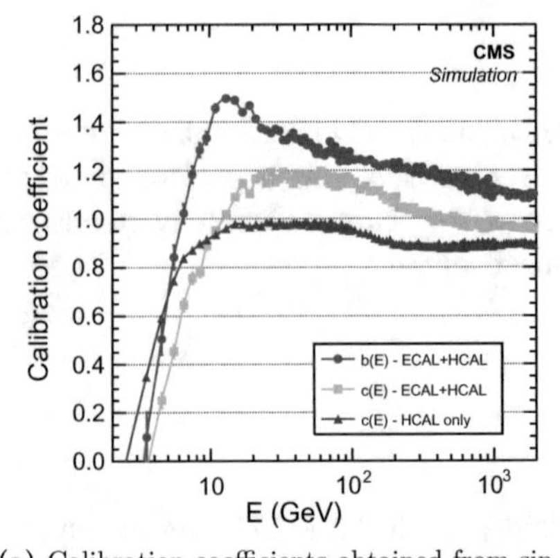

(a) Calibration coefficients obtained from single hadrons as a function of their true energy E. The coefficient for hadrons depositing energy only in the HCAL is shown as triangles. The coefficients for hadrons depositing energy in both the ECAL and HCAL are shown as circles for the ECAL clusters and squares for the HCAL clusters.

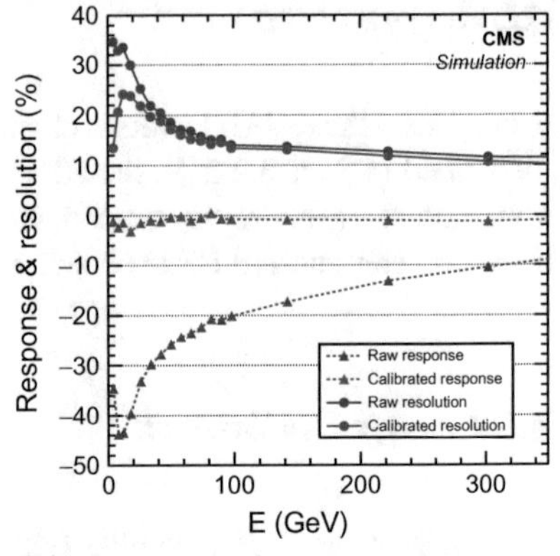

(b) Raw and calibrated energy response (dashed curves) and resolution (full curves) for for single hadrons, as a function of their true energy E. The raw (calibrated) response and and resolution are obtained by a Gaussian fit to the distribution of the relative difference between the raw (calibrated) calorimetric energy and the true hadron energy.

Fig. 3.7 Hadron calibration coefficients and energy response and resolution in the barrel [9]

leave enough energy in the calorimeters to exceed the thresholds of the clustering algorithm. As such, any deposits from these hadrons are due to upward fluctuations of the showering process, which are calibrated away by the small values of b and c. The result is to effectively replace all low energies with a constant a, which is closer to the true hadron energy.

The hadron energy calibration generally affects only 10% of the total measured event energy, which is therefore expected to be modified by only a few percent on average by the calibration procedure.

3.2.4 Muon Detectors

The muon detectors are located outside the solenoid, between and around the three layers of the iron return yoke, and consist of four layers of three different types of gaseous detector planes. Drift tube (DT) chambers are used in the barrel region ($|\eta| < 1.2$) where the neutron background is small, the muon rate is low, and the magnetic field is low. Cathode strip chambers (CSC) are used in the the endcaps ($0.9 < |\eta| < 2.4$), where the muon rate and neutron induced background are large, and the magnetic field is strong. A system of resistive plate chambers (RPC) complement the DT and CSC in the barrel and part of the endcaps, covering the pseudorapidity range $|\eta| < 1.6$. A schematic layout of one quarter of the muon system is shown in Fig. 3.8.

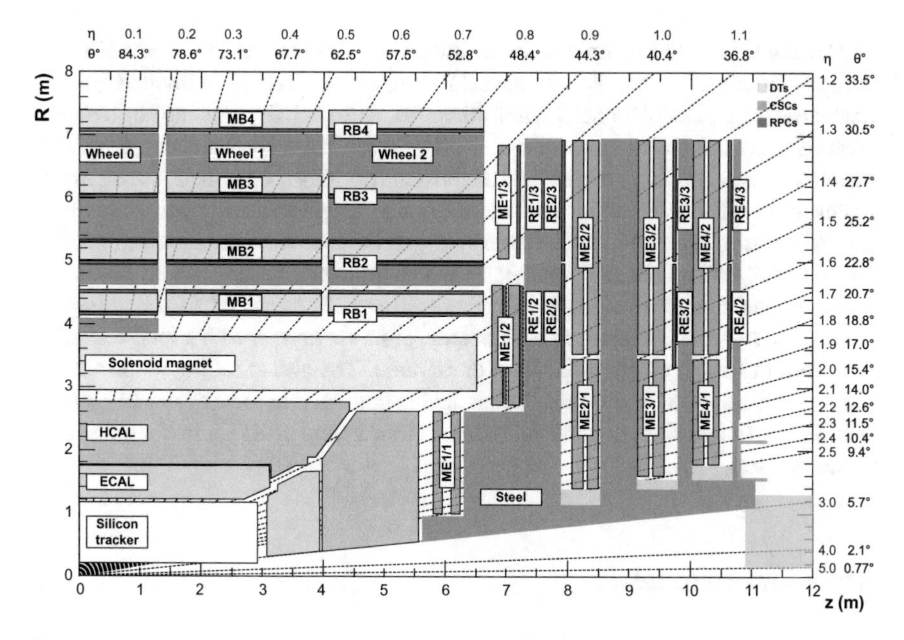

Fig. 3.8 Schematic cross section of one quadrant of the CMS muon system [17]. The DT chambers are shown in the barrel, the CSC in the endcap and the thin RPC in both

The barrel of the muon system includes four layers of DT, called *stations*, at radii of 4.0, 4.9, 5, 9, and 7.0 m from the beam axis, housed on five wheels. Each wheel is made from 12 sectors covering a 30° angle in ϕ, resulting in 48 "positions" per wheel. The chambers in different stations are staggered so that a high-p_{T} muon passing near a sector boundary crosses at least three of the four stations. The top and bottom sectors of the outermost layer host two chambers each, while each other sector and layer hosts a single chamber, thus resulting in 50 DT chambers per wheel. Each DT chamber in the three innermost stations (MB1, MB2 and MB3 in Fig. 3.8) contains 12 planes of aluminium drift tubes spanning about 28 cm: 4 planes measuring r-ϕ, then 4 planes measuring z, then a spacer, and then another 4 planes measuring r-ϕ. The outermost station (MB4) does not contain the z-measuring planes. The maximum drift length is 2.0 cm and the single-point resolution of a plane is approximately 200 μm. Each DT chamber provides a muon vector in space, with a precision in ϕ better than 100 μm or 1 mrad.

Each DT chamber has an RPC attached to its innermost face and the two inner layers have an additional RPC attached to the outermost side. The RPCs are gaseous parallel-plate detectors with pick-up strips sitting between two sets of anode-cathode Bakelite plates, thus forming double-gap modules. They provide a fast response with good time resolution but a coarser position resolution than the DTs or CSCs. The time resolution is sufficiently good to allow the RPCs to unambiguously identify the correct bunch crossing. A high-p_{T} muon in the barrel, having a nearly straight trajectory, will cross up to six RPCs and four DT chambers, producing up to 44 hits in the DT system resulting in a muon-track candidate.

The endcaps of the muon system comprise four stations of CSCs labelled ME1 to ME4, mounted on the disks of the return yoke. Each station is divided into two concentric rings, except for ME1 which has three, with each ring hosting 36 chambers, except for the innermost rings of ME2–ME4 which have 18 chambers. Each CSC chamber consists of seven trapezoidal panels sandwiching six gas gaps, each gap with a plane of radial cathode strips and a plane of anode wires running perpendicular to the central strip. A charged particle traversing a chamber will cause ionisation of the gas and a subsequent electron avalanche, which produces a charge on an anode wire and an image charge on a group of cathode strips. Therefore a CSC chamber will provide the r-ϕ-z coordinates of a hit in each of its six layers, with a total resolution in ϕ from the strips of about 200 μm or 10 mrad. The signal on the wires alone is fast, but has a coarse position resolution, and is thus used in the Level-1 trigger (see Sect. 3.3.1). In addition, the endcap includes four layers of RPCs in the outer rings of each station.

Muon System Readout

The DT readout starts with the trigger electronics and readout board (ROB) mounted in the space inside each chamber. From there both trigger and digital data signals are sent to a sector collector board and readout server board, respectively, located in the detector cavern. The trigger data is then sent to the regional muon trigger, while the full readout data are sent to one of five detector dependant units (DDUs) located in the service cavern, which then send them to the central DAQ system.

There are two readout paths for the CSCs. The anode wire data from each plane is collected in an anode front-end board (FEB) and sent to a trigger board, located on the face of each chamber, which looks for tracks from the six wire hits that point back toward the vertex and send its results to a trigger mother board (TMB) located in one of 60 crates around the edge of the flux-return-yoke disks. The cathode strip pulse heights from each plane are collected in a cathode FEB and sent directly to a trigger logic located on the TMB, which looks for hit patterns in the six cathode strip layers of a chamber. The TMB attempts to match tracks from the cathode strips and anode wires of a chamber and sends its results to a muon port card in the same crate for triggering. The TMB also send the full anode and cathode raw data to the DAQ motherboard (DMB) located in the same crate, which digitises and buffers the data and sends them via optical fibres to a DDU located in the service cavern. Each DDU combines and checks the data from 13 DMBs, and sends it to a data concentration card, which merges the data from nine DDUs and sends them to the DAQ.

The analogue RPC signals are discriminated in the FEBs located on the chambers and then sent to link boards (LBs) located in the detector cavern. Each LB synchronises the signals to the 40 MHz clock and applies zero suppression to compress the data. The information from up to three LBs is multiplexed and converted to optical signals before being transmitted via optical fibres to the trigger boards (TBs) in the service cavern. Each TB deserialises the data and transmits them in parallel to the RPC trigger pattern comparator and the readout mezzanine boards (RMB), which demultiplex the data and buffers them, awaiting a trigger signal. Three data concentration cards then collect the data from the RMBs and send them to the DAQ.

Muon Track Reconstruction

Muon tracks can be reconstructed from the muon system alone, by matching hits from all DT, CSC and RPC planes. The efficiency and precision of the measured p_{T} can be greatly enhanced by including information from the tracker. The muon reconstruction algorithms are described in Sect. 4.3.3.

3.3 The CMS Trigger System

At the design centre-of-mass energy and luminosity of the LHC, 13 TeV and $10^{34}\,\mathrm{cm}^{-2}\mathrm{s}^{-1}$ respectively, with a proton-proton cross section of $\approx 70\,\mathrm{mb}$, there are expected to be around 700 million p-p collisions per second, corresponding to an event rate of $\approx 700\,\mathrm{MHz}$, which are generally accumulated in time around the 40 MHz bunch-crossing rate. For an accurate measurement of the event, the full information from all subdetectors must be collected and pieced together to form a complete image of all particles produced in the collision. Collecting this information in a given time window will also capture all collisions in that window. The term *event* is used to refer to all collisions and corresponding physics processes occurring in the given time interval.

Saving the information read out from all the subdetectors for each event is both impractical and impossible. It is impractical because the vast majority of these events are uninteresting in terms of physics, since they do not produce new or little-known particles, and having to sort through all these events at a later time, looking for the tiny fraction of interesting events, would be time consuming and resource intensive. More importantly, it is impossible for two main reasons. First, the readout of the detectors is not fast enough to allow the full detector information to be stored for each bunch crossing—although the detector response is matched to the nominal bunch-crossing frequency of 40 MHz, the readout of some subsystems is much slower. Second, since each event stored for later analysis requires about 1 MB of disk space, the write out and storage of event information is not fast enough—it would require a write-out speed of $\sim$40 TB/s.

To select events of interest for storage and later analysis, CMS uses a two-tiered trigger system. The first level (L1) is composed of custom hardware and uses partial, fast-response data from the calorimeters and muon system to identify and select events containing candidate objects, i.e. muons, electrons, photons or jets, at a rate of up to 100 kHz within 4 μs. The second level, known as the high-level trigger (HLT), runs a version of the full event reconstruction software optimised for fast processing on a farm of commercial processors, and reduces the event rate to around 1 kHz for storage to disk. Further details of the two trigger levels are given below, while a full description is provided in Ref. [18].

As stated in Sect. 3.1, multiple events can occur at each bunch crossing, i.e. within a 25 ns window. Capturing the full detector information for one event, will also capture other events occurring at the same or very near time.

3.3.1 Level 1 Trigger

The L1 trigger is a hardware-based system with a fixed latency. It has a maximum output rate of 100 kHz and maximum processing time of 4 μs per collision. Within these 4 μs, the L1 system must decide if an event should be accepted for further processing or permanently rejected, using partial information from the muon detectors and calorimeters. Specifically, it looks for ionisation deposits in the DT, CSC and RPC that are consistent with a muon, and energy clusters in the ECAL, HCAL and HF that are consistent with an electron, photon, hadron jet, τ-lepton jet, missing transverse momentum ($p_{\mathrm{T}}^{\mathrm{miss}}$), or a large scalar sum of jet transverse momenta (H_{T}). The final trigger decision is based on a programmable menu which uses these candidate objects to assess if any of up to 128 selection algorithms are satisfied.

A schematic diagram of the L1 trigger logic is shown in Fig. 3.9. The trigger primitives from the calorimeters and muon detectors are processed in several steps before the combined event information is evaluated in the global trigger (GT), which decides whether to accept the event or not. Accepted events are then passed on to the HLT, described in Sect. 3.3.2, via the DAQ, described in Sect. 3.4, for further processing, selection and eventual storage. To mitigate the data losses due to hardware failures, spare electronics modules for all systems of the L1 trigger are kept in the service cavern, and an entire replica of the GT is kept running, ready to take over at any time.

The L1 Calorimeter Trigger

The calorimeter-based component of the L1 trigger consists of two stages: a regional calorimeter trigger (RTC) and a global calorimeter trigger (GCT). The RCT receives the energy measurements and quality flags from over 8 000 ECAL crystals and HCAL and HF towers, covering the region $|\eta| < 5$. It then processes this information in parallel to determine electron and photon candidates and regional E_{T} sums based on blocks of 4×4 cells. The GCT further processes the e/γ candidates, identifies and classifies jets, as central, forward, and tau jets, using the E_{T} sums, and calculates global quantities such as $p_{\mathrm{T}}^{\mathrm{miss}}$ and H_{T}. Its output includes two types of electron and photon (isolated and nonisolated), four types each of central, forward and tau jets, and several global quantities.

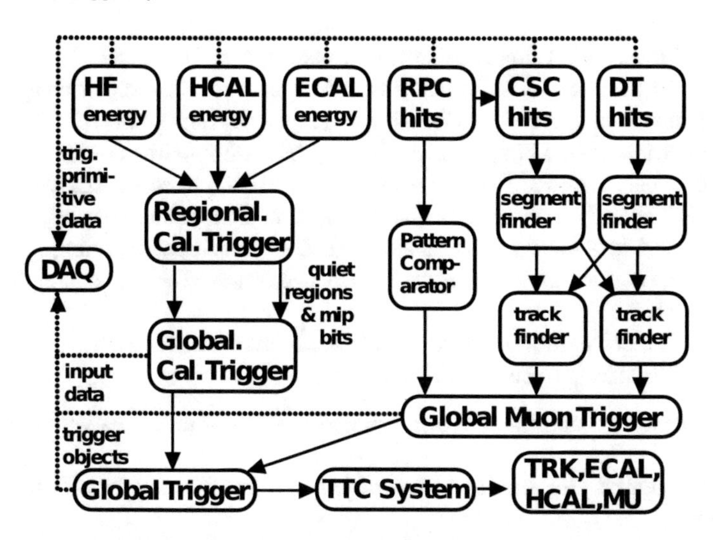

Fig. 3.9 Overview of the CMS L1 trigger system [18]. Data from the HF, HCAL and ECAL are first processed in the regional calorimeter trigger (RCT) and then in the global calorimeter trigger (GCT). Hits from the RPC are processed via a pattern comparator, while those from the CSC and DT are processed in a system of segment- and track-finders, before being sent onwards to the global muon trigger (GMT). The results from the GCT and GMT are combined in the global trigger (GT), which makes the final trigger decision. The decision is sent to the tracker (TRK), ECAL, HCAL and muon systems (MU) via the trigger, timing and control (TTC) system for full read out. The DAQ reads data from various subsystems (not all links shown) for further processing and offline storage. The acronym "mip" stands for minimum-ionising particle

The basic calorimeter blocks for the L1 trigger are trigger towers, which correspond to the 5×5 crystal towers defined in Sect. 3.2.2 for the ECAL barrel. In the ECAL endcap however, the trigger towers are collections of groups of five contiguous crystals and may extend over more than one 5×5 supercrystal. The transverse energy deposited in the crystals of a trigger tower is summed to create a trigger primitive (TP). In the barrel, the TPs are calculated by the front-end electronics and sent to the off-detector trigger concentrator cards (TCCs), while in the endcaps the TPs are calculated in the TCCs. Data from the TCCs are then sent to the RCT.

In the HCAL the TPs are computed by the HTR, and include the data from a single readout (clock period, or bunch crossing) in the barrel, two readouts in the endcaps, and up to 12 readouts in the HF. An important task of the TP generation is to assign the correct bunch crossing to the detector pulses, which can span over several clock periods. This is accomplished by digitally filtering the energy samples and applying a peak finder algorithm. The sum of the amplitudes of the maximum peak and the following time period are used to estimate the pulse energy, while the position of the peak determines the timing. The E_{T} of each HCAL trigger tower is calculated on a linear scale of 10 bits, where an overflow is set to the scale maximum. In the TP, this 10-bit energy is converted to an 8-bit nonlinear scale to reduce the data flow before transmission to the RCT.

Each of the 18 RCT crates collects information from the ECAL, HCAL and HF towers in 24-bit packets, comprising two 8-bit energies (the electromagnetic and hadron components), either two ECAL fine-grain (FG) bits or two HCAL minimum ionising particle (MIP) bits, a bunch crossing bit, and five bits of error codes. In a series of stages, each RCT processes the information and sends out the following to the GCT: the top four isolated and the top four non-isolated e/γ candidates, 14 regional 4×4 tower sums of ECAL+HCAL E_{T}, with a corresponding quiet bit for muon isolation, a τ-veto bit, and the logical OR of the MIP bits, and eight HF TP and quality bits.

The GCT receives the output from the 18 RCT crates and, for each event, computes the following objects to be sent to the GT:

- the four isolated and the four non isolated photons or electrons with the highest E_{T};
- the four central jets, the four forward jets and the four τ jets with the highest energy;
- the total transverse energy, $S_{\mathrm{T}} = \sum E_{\mathrm{T}}$; the H_{T}; the $p_{\mathrm{T}}^{\mathrm{miss}}$; the missing jet transverse energy; and
- the sum of fine-grain bits and the sum of transverse energies in the HF.

It computes all the jet related quantities within 24 bunch crossings (600 ns) and the electrons/photons within 15 bunch crossings (375 ns). The GCT also acts as a readout device for itself and the RCT by storing information until it receives an L1 accept and then sending it to the DAQ.

The L1 Muon Trigger

The L1 muon-trigger system uses information from all three muon detectors. The data from the DT, CSC and RPC are processed in a number of stages to build the final muon candidates.

Local muon track segments for the trigger (primitives) are formed on the DT and CSC front end. Local DT electronics reconstruct track segments and compute the radial position, the bending angle, the number of layers used and hits along the longitudinal direction, with a 94% efficiency on the bunch crossing identification. The DT trigger segments are sent over a 6 Gb/s optical link to the DT track finder (DTTF). In the case of the CSC, local charged-track segments are constructed separately from the cathode and anode hits, and correlated in the trigger motherboard (TMB). The azimuthal position and radial distance of a track as well as precise timing information are sent over an optical-fibre link to the CSC track finder (CSCTF).

The DTTF uses the information from the local DT trigger of each station to reconstruct muon candidates and determine their p_{T}. For each of the three inner stations, it computes, via a look-up table, the expected position at the outer stations, while for the outermost station the extrapolation is done inwards. It then compares the actual segments to the expected positions and accepts them as track segments if they fall within a programmable tolerance, to build a track candidate. It uses the difference in the azimuthal positions of the two innermost segments to compute the p_{T} of the track. The tracks from different regions of the detector are sorted based on reconstruction quality and p_{T} and the best four are sent to the global muon trigger (GMT).

The CSCTF performs a pairwise comparison of track segments in different stations, testing compatibility of ϕ and η, and accepts a match if they fall within a programmable tolerance. Matched segments are used to build tracks of at least two stations and calculate their p_{T}. The track finder can accept segments in different bunch crossings, by considering a sliding time window and cancelling duplicate tracks. The reported bunch crossing is given by the second earliest track segment. The CSCTF also identifies and triggers on beam halo muons, arising from proton interactions with gas in the beam pipe or accelerator material, for monitoring and veto purposes. Like the DTTF, the CSCTF sorts tracks from different regions of the detector based on reconstruction quality and p_{T} and sends the best four to the GMT.

The RPC provides a dedicated and complementary trigger system with excellent time resolution of the order of 1 ns, to determine the correct beam-crossing time at high luminosities. Unlike the CSC and DT, the RPC does not form trigger primitives, but the spatial and temporal coincidence of hits in its different layers are used directly to reconstruct muon trigger candidates. The pattern comparator trigger (PACT) compares signals from all RPC layers to predefined hit patterns in order to find muon candidates and assigns the muon p_{T}, charge, η, and ϕ to the matched pattern. The trigger algorithm requires a minimum number of hits depending on the position of the muon, with typically at least 3 or 4 hits required. After a system wide sorting of muon candidates the four best candidates from the barrel and 4 best candidates from the endcaps are sent to GMT for subtrigger merging.

The GMT performs a number of functions based on the information it receives from the DTTF, CSCTF, and RPC trigger systems. It synchronises incoming regional muon candidates, merges or cancels duplicate candidates, assigns an optimised p_{T} to merged candidates, sorts candidates according to programmable criteria, assigns quality codes to outgoing candidates and stores information about the incoming and outgoing muon candidates in the event data. Most of the GMT logic is implemented in the form of look-up tables, which enables a high level of flexibility and functional adaptability without having to change the FPGA[9] firmware, e.g. to adjust selection requirements, such as p_{T}, η, and quality, of the regional muon candidates. The final stage of processing involves the sorting of muon candidates according to the ranking criteria, first separately in the barrel and endcap regions to determine the best four candidates in each, and then globally to send the four highest ranked candidates to the GT.

[9] A field-programmable gate array (FPGA) is an integrated circuit that can be configured by the end user.

The L1 Global Trigger

The GT is the final stage of the L1 trigger system, consisting of several VME boards hosting FPGAs, located in a single crate in the service cavern. For each bunch crossing, it uses the incoming trigger objects from the GCT and GMT to decide whether to accept an event for subsequent evaluation by the HLT, or permanently reject it.

The L1 trigger menu is implemented in firmware and consists of up to 128 *algorithm triggers* and up to 64 *technical triggers*. The algorithm triggers combine conditions on trigger objects, e.g. electron $p_T > 20\,\text{GeV}$, in a simple AND-OR-NOT logic for later use in the HLT. The technical triggers use special signals directly from the subsystems to trigger or veto the decision, and are used for monitoring and calibration of the subdetectors and the L1 trigger system itself. The L1 menu also contains an array of prescale factors for each trigger, which determine how often the trigger is active in order to reduce its output rate.[10] The algorithm bits are combined in a final OR, such that an accept from any of them triggers the readout of the whole CMS detector and the transmission of all data to the HLT for further evaluation. The L1 accept signal can be blocked by trigger rules, i.e. programmable criteria to limit the readout frequency of certain subdetectors, or detector deadtime, e.g. busy signals from subdetectors.

An example of an L1 trigger menu used in 2016 is shown in Table 3.2. It shows a selection of algorithm and technical triggers, the conditions of which are evident from the names, along with their prescale factors. A menu such as this can be modified frequently (up to several times per day) during commissioning and testing, and is otherwise kept for several days, weeks or months during stable running. The modified L1 trigger menus are implemented by loading another firmware version to the GT, and reconfiguring it. The choice of prescale "column" is configurable and can be modified during operation, without reloading the firmware. The GT system logs all trigger rates and deadtimes, which are monitored live to ensure smooth operation.

Beam Position Timing Trigger System

The LHC operates beam position monitors around the LHC ring. The closest two for each interaction point are reserved for timing measurements and are named beam pick-up timing experiment (BPTX) detectors. For CMS, they are located at $Z \approx \pm 157\,\text{m}$ and referred to as `BPTX+` and `BPTX-`. A dedicated trigger determines valid bunch crossings by requiring a coincidence between the monitors on each side, i.e. `BPTX_AND = BPTX+ AND BPTX-`. In some cases, low threshold triggers subject to high background noise that would normally render them unusable, can be successfully deployed by requiring a coincidence with `BPTX_AND`.

---

[10] For example, a prescale of 10 means that the trigger will only assess one out of every 10 events. This allows low threshold paths, which would normally have excessively high rates, to select some events that would otherwise be lost and is mainly used for calibration, efficiency measurements and testing.

Table 3.2 Excerpts of an L1 trigger menu used for a brief period in May 2016 for data taking [19]. The prescale column can be chosen to target a particular instantaneous luminosity, or for an emergency situation (e.g. faulty configuration or extremely high rates). The suffix "er" stands for η-restricted. HTT is the H_T calculated by the calorimeter trigger, ETM is the p_T^{miss}, Tau is a τ jet, and EG is an electron or photon (e/γ)

Menu name: L1Menu_Collisions2016_v2c

Bit	Name	Prescale column							
		Emerg	1e34	7e33	5e33	3.5e33	2e33	1e33	...
--- Selected algorithm triggers ---									
0	L1_ZeroBias	347	347	347	347	347	347	347	...
6	L1_SingleMu12	0	900	450	360	270	180	90	...
7	L1_SingleMu14	0	40	30	20	16	8	4	...
8	L1_SingleMu16	0	40	30	20	16	8	1	...
9	L1_SingleMu18	0	1	1	1	1	1	1	...
13	L1_SingleMu30	0	1	1	1	1	1	1	...
13	L1_SingleMu30	0	1	1	1	1	1	1	...
17	L1_SingleMu16er	0	1	1	1	1	1	1	...
27	L1_DoubleMu_11_4	0	1	1	1	1	1	1	...
37	L1_TripleMu_5_5_3	0	1	1	1	1	1	1	...
44	L1_SingleEG26	0	1500	1000	700	500	300	1	...
45	L1_SingleEG28	0	1	1	1	1	1	1	...
53	L1_SingleIsoEG18	0	1000	700	500	300	200	1	...
54	L1_SingleIsoEG20	0	1	1	1	1	1	1	...
62	L1_SingleIsoEG18er	0	1	1	1	1	1	1	...
71	L1_DoubleEG_15_10	0	1	1	1	1	1	1	...
84	L1_SingleJet90	0	3500	2450	1750	1400	700	1	...
85	L1_SingleJet120	0	1	1	1	1	1	1	...
94	L1_DoubleJetC60	0	400	300	200	120	80	1	...
95	L1_DoubleJetC80	0	1	1	1	1	1	1	...
102	L1_QuadJetC40	0	1	1	1	1	1	1	...
105	L1_SingleTau80er	0	1	1	1	1	1	1	...
109	L1_DoubleIsoTau28er	0	1	1	1	1	1	1	...
118	L1_HTT220	0	6000	4000	3000	2000	1200	50	...
119	L1_HTT240	0	1	1	1	1	1	1	...
125	L1_ETM80	0	1	1	1	1	1	1	...
130	L1_HTM100	0	100	70	50	35	20	10	...
144	L1_Mu5_EG15	0	1	1	1	1	1	1	...
154	L1_Mu16er_Tau20er	0	1	1	1	1	1	1	...
168	L1_Mu6_DoubleEG17	0	1	1	1	1	1	1	...
172	L1_Mu6_HTT200	0	1	1	1	1	1	1	...
177	L1_QuadJetC36_Tau52	0	1	1	1	1	1	1	...
--- Selected technical triggers ---									
219	L1_IsolatedBunch	0	23	23	23	23	23	23	...
221	L1_BeamGasPlus	0	10	10	10	10	10	10	...
233	L1_BPTX_TRIG2_AND	0	0	0	0	0	0	0	...

3.3.2 High Level Trigger

The selection of interesting events at the HLT requires all objects in an event to be reconstructed, which is performed similarly to the reconstruction used in offline processing. The objects include electrons, photons, muons, and jets, reconstructed using the full detector information from all subsystems.

The HLT hardware consists of a processor farm of commercially available computers, referred to as the event filter farm (EVF), and is located on the surface of the CMS site. The EVF runs Scientific Linux and is structured in many blocks of filter-builder units, one of each was originally installed on a single multi-core machine communicating via shared memory. With the Run-2 upgrade, these units were separated to different machines and connected via 1-10-40 GB/s Ethernet links. Each builder unit (BU) assembles complete events from individual fragments received from the subdetectors, and then sends it to specific filter units (FUs) upon request. The filter units then unpack the raw data into detector-specific data structures and execute the object reconstruction and trigger filtering. In total there were around 22 000 CPU cores in 2016. Given the maximum L1 input rate of 100 kHz and the number of EVF cores, implies that the average processing time for events in the HLT cannot exceed around 220 ms. Since the time required for the reconstruction of a full event and subsequent filtering can be up to 2 s, most events must be rejected or accepted quickly, with only partial event information.

The data processing of the HLT is performed in a number of *HLT paths*, each of which is a set of processing algorithms that both reconstructs and makes selections on physics objects. Each HLT path is implemented in a predefined sequence of steps, starting with an L1 trigger *seed* consisting of one or more L1 triggers, and then increasing in complexity, such that events failing any step are immediately rejected. Information from the calorimeters and muon detectors are used first in early steps, before the CPU-intensive tracker reconstruction is performed. If an event successfully passes the final step of an HLT path it is immediately accepted. Many HLT paths are executed on a single event in parallel. As soon as one path accepts the event, the EVF processing stops and the full event information is written out to disk. On the other hand, an event is rejected if all HLT paths reject the event. The processing benefits from the sharing of objects, as an object only needs to be reconstructed once for use in several HLT paths, as well as some fast paths that do not run the tracker reconstruction at all. For any reconstruction beyond calorimeter and muon detector objects, the particle flow algorithm described in Sect. 4.3.1 is executed, which includes the full reconstruction of the tracker.

Upon successful acceptance by the HLT, event data are stored locally on disk and then enter a transfer queue to the CMS Tier-0 computing centre for offline processing and permanent storage. During the offline processing, the full event reconstruction is completed and events are grouped into a set of non-exclusive "streams" based on the types of HLT paths which have accepted the events. The total output rate of the HLT is limited by the size of the events and the rate at which the CMS Tier-0 can process events. The typical size of a fully reconstructed event is approximately 1 MB. In addition to the primary event stream for physics analysis, monitoring and calibration streams are also recorded. These streams usually save events with reduced content of just a few kB or are selected by triggers with large prescale factors, to avoid saturating the data taking bandwidth. In 2016, the maximum sustainable HLT output rate was slightly above 1 kHz, while the peak rate could be as high as 2 kHz.

Similar to the L1 trigger, HLT paths are grouped together in an HLT menu, which is uploaded to the DAQ system. There were around 500 paths on the HLT menu in 2016, most of which were developed for physics analysis. Most paths are developed and maintained by the physics object groups for use in several analyses. These paths typically contain selections on one or more common objects such as muons, electrons, photons and jets, but their complexity is kept low. Many other paths however are developed for a specific analysis, such as the paths described in Sect. 4.2 for use in this analysis. These paths often contain complex mixtures of objects and selections targeting a specific final state of a specific physics process. As for the L1 menu, an array of prescale factors targeting certain instantaneous luminosities accompanies each HLT path, such that the final HLT output rate is kept near its sustainable maximum as the luminosity decreases throughout a fill.

An excerpt from an example HLT menu used in 2016 data taking is shown in Table 3.3. The HLT menu is under constant modification with new paths being developed by end users as required. The final responsibility of the HLT menu lies with the trigger coordination, which must approve each path that is modified or added to the menu. Once a new menu is created, it can be uploaded to the DAQ and activated with a reconfiguration. This is done in the CMS control room, by the HLT expert-on-call and the DAQ shifter.

3.3.3 Trigger Maintenance

The development and maintenance of the CMS trigger system is the responsibility of the trigger coordination, which is divided into four subgroups. The L1 subgroup deals with the development of the L1 trigger and the L1 menu. For the HLT, there are three subgroups:

- Strategy Trigger Evaluation and Monitoring (STEAM) is responsible for:
  - estimating and measuring the rates of individual HLT paths and the full HLT menu, for additions to the menu and adjustments to prescales;
  - validating HLT menus and maintaining the trigger-related data-quality monitoring (DQM) software;
  - analysing and reporting the HLT performance; and
  - producing Monte Carlo simulation event samples for use in trigger rate estimation, testing and calibration.
- Software Tools Online Release Menu (STORM) is responsible for:
  - integrating new paths and all modifications to the HLT menu;
  - developing the HLT development framework and tools; and
  - maintaining the HLT menu database software.

Table 3.3 Selected paths from an HLT menu used for a brief period in September 2016 for data taking [20]. The prescale column can be chosen to target a particular instantaneous luminosity, or for an emergency situation (e.g. faulty configuration or extremely high rates)

Menu name: /cdaq/physics/Run2016/25ns15e33/v4.1.4/HLT/V1

HLT Path Name	Prescale column							
	Emerg	1.25e34	1.15e34	1.05e34	9.5e33	8.5e33	7.5e33	...
- - - Stream: Calibration - - -								
HLT_EcalCalibration_v3	1	1	1	1	1	1	1	...
- - - Stream: PhysicsEGammaCommissioning - - -								
HLT_L1FatEvents_v1	70	70	70	70	70	70	70	...
HLT_Ele25_WPTight_Gsf_v7	0	0	0	0	0	1	1	...
HLT_Ele27_WPTight_Gsf_v7	1	1	1	1	1	1	1	...
HLT_Photon120_v7	160	145	130	115	100	80	70	...
HLT_Photon175_v8	1	1	1	1	1	1	1	...
HLT_Ele17_Ele12_CaloIdL_TrackIdL_IsoVL_DZ_v9	0	0	0	0	0	1	1	...
HLT_Ele23_Ele12_CaloIdL_TrackIdL_IsoVL_DZ_v9	1	1	1	1	1	1	1	...
HLT_Ele23_Ele12_CaloIdL_TrackIdL_IsoVL_v9	15	13	12	11	10	9	8	...
HLT_JetE30_NoBPTX_v4	1	1	1	1	1	1	1	...
- - - Stream: PhysicsHadronsTaus - - -								
HLT_QuadJet45_DoubleBTagCSV_p087_v6	83	76	68	60	52	42	37	...
HLT_QuadJet45_TripleBTagCSV_p087_v6	1	1	1	1	1	1	1	...
HLT_BTagMu_DiJet20_Mu5_v5	7	6	6	5	6	6	6	...
HLT_BTagMu_DiJet40_Mu5_v5	1	1	1	1	1	1	1	...
HLT_PFHT300_PFMET110_v6	1	1	1	1	1	1	1	...

(continued)

Table 3.3 (continued)

Menu name: /cdaq/physics/Run2016/25ns15e33/v4.1.4/HLT/V1

HLT Path Name	Prescale column							
	Emerg	1.25e34	1.15e34	1.05e34	9.5e33	8.5e33	7.5e33	...
HLT_DiPFJetAve260_v8	140	126	113	100	87	70	61	...
HLT_DiPFJetAve300_HFJEC_v9	1	1	1	1	1	1	1	...
HLT_PFHT650_v8	132	120	105	95	85	65	58	...
HLT_PFHT900_v6	1	1	1	1	1	1	1	...
HLT_PFJet400_v9	18	15	14	12	10	8	7	...
HLT_PFJet450_v9	1	1	1	1	1	1	1	...
HLT_PFHT400_SixJet30_DoubleBTagCSV_p056_v5	1	1	1	1	1	1	1	...
HLT_PFHT400_SixJet30_v7	130	110	100	90	80	65	55	...
HLT_LooseIsoPFTau50_Trk30_eta2p1_MET110_v5	1	1	1	1	1	1	1	...
HLT_DiPFJet40_DEta3p5_MJJ600_PFMETNoMu140_v5	1	1	1	1	1	1	1	...
HLT_MET200_v4	1	1	1	1	1	1	1	...
HLT_PFMET300_v6	1	1	1	1	1	1	1	...
- - - Stream: PhysicsMuons - - -								
HLT_IsoMu22_v5	0	0	0	0	1	1	1	...
HLT_IsoMu24_v4	1	1	1	1	1	1	1	...
HLT_IsoTkMu22_v5	0	0	0	0	1	1	1	...
HLT_IsoTkMu24_eta2p1_v1	1	1	1	1	1	1	1	...
HLT_IsoMu19_eta2p1_LooseCombinedIsoPFTau20_v1	1	1	1	1	1	1	1	...
HLT_Mu17_TrkIsoVVL_Mu8_TrkIsoVVL_v6	42	38	34	30	26	21	18	...
HLT_Mu17_TrkIsoVVL_TkMu8_TrkIsoVVL_DZ_v6	1	1	1	1	1	1	1	...
HLT_Mu23_TrkIsoVVL_Ele12_CaloIdL_TrackIdL_IsoVL_DZ_v4	1	1	1	1	1	1	1	...
HLT_Mu23_TrkIsoVVL_Ele12_CaloIdL_TrackIdL_IsoVL_v9	6	6	5	5	4	4	3	...

- Field Operation Group (FOG) is responsible for:
  - monitoring the usage and performance of the EVF;
  - monitoring the live trigger rates and developing the software for such monitoring; and
  - ensuring smooth integration with the DAQ.

In addition to these subgroups, end users are responsible for the management of individual HLT paths. This is organised through a subgroup of each physics analysis group (responsible for analysing data) or physics object group (responsible for the reconstruction of a particle or object) dedicated to trigger development, validation and monitoring.

A separate branch of trigger coordination is responsible for the live operations of the trigger. It essentially falls under the CMS run coordination, which ensures the smooth operation of the detector. For the trigger, there is a dedicated L1 trigger shifter in the CMS control room at all times during operation. The trigger shifter is responsible for monitoring the instantaneous L1 and HLT trigger rates, adjusting the prescale column as necessary and reconfiguring the trigger for firmware updates or other reasons. An L1 expert-on-call is responsible for making changes to the live L1 trigger menu and uploading new firmware. Similarly, an HLT expert-on-call is responsible for the changes to the HLT menu and uploading new configurations to the DAQ.

During my work at CMS, I made original contributions to the FOG subgroup through the development of a trigger rate monitoring package that provides live rates to the DAQ system. I also contributed to the STEAM subgroup by performing rate estimations and was the responsible for producing simulated event samples for two years.

Rates Estimation and Measurement

The live L1 and HLT trigger rates are provided by the online rate monitoring system and observed in the CMS control room. The online monitoring software provides the total L1 rate as well as rates for each L1 trigger, while for the HLT it provides the total rate and the rate for each stream. The individual HLT path rates are also available through a separate slightly delayed system called *CMS web based monitoring* (WBM). While the rate measurement of the current HLT menu in live data is automatically provided, the rate of a new or modified path must be estimated before inclusion in the online menu.

The rate estimation of new paths is necessary to ensure that it will not push the total HLT rate over sustainable limits. Generally a path will have an allocated bandwidth budget, which it should not exceed. This rate budget is usually expressed as a *unique rate*, i.e. the incremental rate that the path adds to the entire HLT menu, since events accepted by existing triggers do not consume additional bandwidth. Typically, from one data-taking year to the next, a big effort is made to revamp the HLT menu, with end users encouraged to optimise the use of HLT bandwidth, i.e. to ensure they get the best efficiency for their required process for a given rate. Therefore many new paths enter the menu while some paths are slightly modified and others are removed, in an effort to adapt to the expected luminosity of the LHC beams for the coming year.

For each new and modified path, a reliable estimate of its rate is made by running the full HLT reconstruction software on raw events, either from existing data or from simulation. In the case of simulated events, the simulated process must represent the greatest background for the given path. This background is typically from QCD multijet events, for jet and H_{T} based triggers, or leptons from Drell-Yan production or W boson production in association with jets for lepton and photon based triggers. For each simulated process, the total event rate r^{proc} is given by Eq. (3.1), while the rate of a given trigger i is calculated as:

$$r_i^{\mathrm{proc}} = r^{\mathrm{proc}} \cdot \frac{N_i^{\mathrm{proc}}}{N_{\mathrm{tot}}^{\mathrm{proc}}}, \tag{3.9}$$

where N_i^{proc} is the number of events accepted by trigger i and $N_{\mathrm{tot}}^{\mathrm{proc}}$ is the total number of generated events for the process. In the case of assessing the full L1 or HLT menu rate, $r_{\mathrm{menu}}^{\mathrm{proc}}$ is given by Eq. (3.9) by simply replacing N_i^{proc} with the number of events accepted by any trigger in the menu, $N_{\mathrm{menu}}^{\mathrm{proc}}$. The total rate for any given trigger is then the sum of rates from all contributing processes:

$$r_i = \sum_{\mathrm{proc}} r_i^{\mathrm{proc}}. \tag{3.10}$$

To allow the trigger rate to be calculated from data, a special HLT stream is saved which includes all events passing any L1 trigger. The contribution from the L1 zero bias trigger[11] is reduced in this stream to prevent zero bias events dominating the stream. Additionally, to reduce bandwidth consumption, a prescale factor of $\mathcal{O}(100)$ is applied to this special stream. In this way a "full" sample of all events is provided to the HLT menu, for each HLT path to assess, with the idea being that all events that would be accepted by any HLT path, subject to the prescale, are included in the stream. In the case of estimating the rate of a new path, the full event reconstruction is performed with the new path included, such that its trigger decision can be assessed. The estimated rate of an HLT path i is calculated as:

$$R_i = \frac{L_{\mathrm{target}}}{L_{\mathrm{data}}} \cdot F_{\mathrm{PS}} \cdot N_i \cdot \frac{1}{t_{\mathrm{data}}}, \tag{3.11}$$

[11]The zero bias trigger is run with a high prescale $\mathcal{O}(10^3 - 10^4)$ and reads out every event according to its prescale, regardless of whether or not the event is accepted by any object or selection based trigger.

where L_{data} and t_{data} are the luminosity and total time for which the data was recorded, L_{target} is the luminosity at which the rate is to be estimated, F_{PS} is the prescale factor on the HLT stream, and N_i is the number of events accepted by trigger i. The rate of the full menu can be calculated by using N_{menu} in place of N_i in Eq. (3.11).

3.4 The CMS Data Acquisition System

The CMS DAQ system is closely integrated with the trigger system and is responsible for collecting and processing the data from all subdetectors. It is designed to read out the detectors at up to the nominal LHC bunch crossing frequency of 40 MHz, and to sustain a full readout of all detectors at up to maximum L1 trigger rate of 100 kHz. With a full event size of up to 2 MB, the DAQ must therefore sustain an input rate of up to 200 GB/s from approximately 700 different data sources. The DAQ system includes the EVF of the HLT, which reduces the input rate down to about 1 kHz, which then becomes the DAQ output rate to the storage system. A general overview of the DAQ architecture is shown in Fig. 3.10. An overview of the CMS DAQ system is provided in the following, while full details of the original DAQ design can be found in Refs. [8, 21], and details of the upgraded system used in Run-2 are provided in Ref. [22].

3.4.1 DAQ Infrastructure

A full overview of the DAQ architecture is shown in Fig. 3.11. An explanation of the DAQ system, in the order of data flow, from top to bottom of this diagram, is given below.

The DAQ process begins with the readout from the front end drivers (FEDs) of the subdetectors. The detector information is collected in fragments of up to 4 kB for

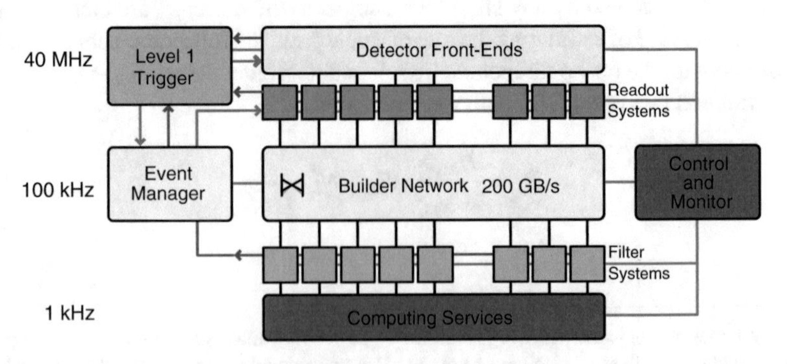

Fig. 3.10 Schematic overview of the CMS DAQ system [8]

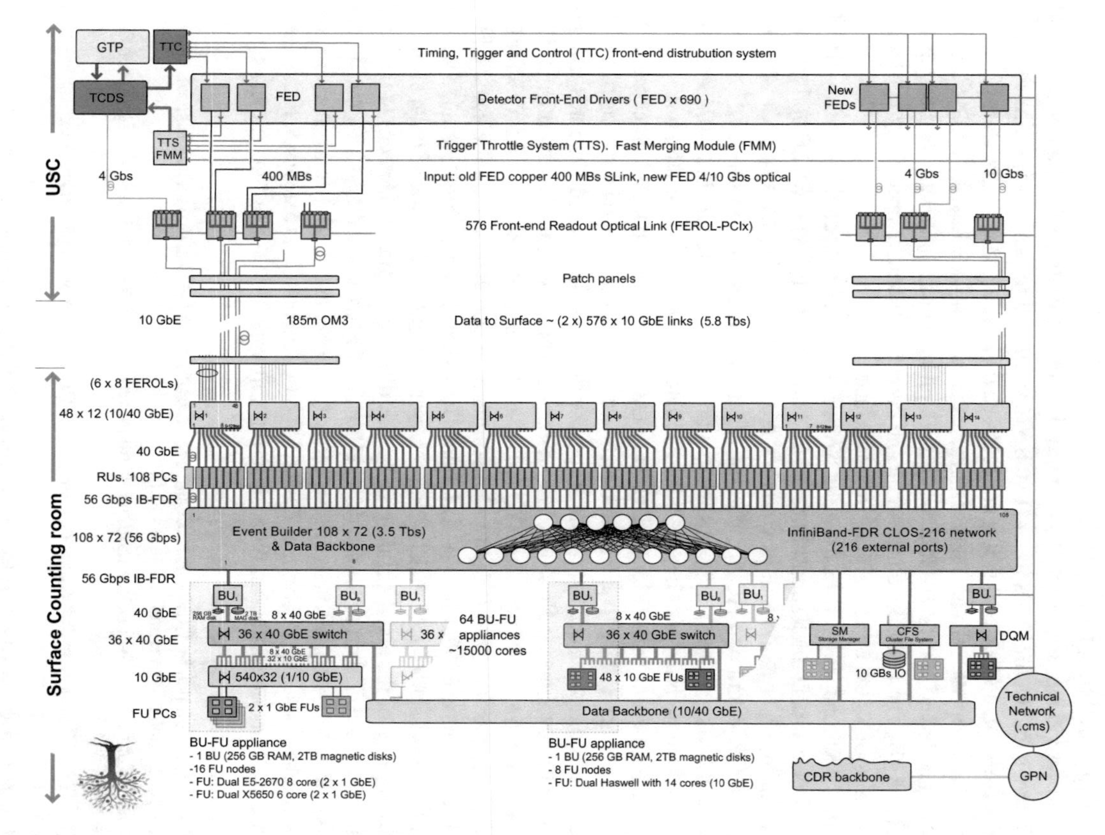

Fig. 3.11 Detailed overview of the CMS DAQ system [22], explicitly showing the link types. A description of the diagram and definition of the acronyms are provided in the text

older readout electronics with S-LINK64,[12] and up to 8 kB for new μTCA[13] based readout electronics with S-LINK Express. The event building is performed by the DAQ in two stages, first by electronics located in the underground service cavern, and then by computing systems located on the surface. The information transfer from underground to the surface is made via 576 individual 10 Gb/s Ethernet (GbE) links, allowing a total transfer rate of 5.8 Tb/s.

The first stage of the DAQ is a pure data concentrator. Information from 690 different subdetector FEDs is sent to the 576 front-end readout optical link PCI-X[14] cards (FEROL-PCIx). 640 FEDs are connected to the FEROL-PCIx with 400 Mb/s S-LINK64 copper links and send fragments from 1 to 4 kB in size, while 50 FEDs are connected with 4 or 10 Gb/s S-LINK Express optical links and send fragments of 2 to 8 kB. The FEROLs concentrate and convert the data to optical signals before transmission to the event builder on the surface.

In the second stage, data from the 576 FEROLs are collected by 14 10-40 GbE switches, each of which streamlines the data and transmits it via 40 GbE optical links to seven or eight readout units (RUs). The 108 RUs are commercial PCs with two 12-core CPUs and two NCIs,[15] one for Ethernet and one for InfiniBand,[16] running standard Linux sockets, and can sustain a throughput of 40 Gb/s. Each RU sends its data to the core builder unit via a 56 Gb/s FDR InfiniBand[17] (IB-FDR) link.

The core event builder takes data from the 108 RUs as input and processes it on 72 builder units (BUs). The RUs and BUs are all inter-connected by means of a Clos network[18] composed of 30 separate 36-port switches, 18 leaves and 12 spines, a larger version of that shown in Fig. 3.12. The switching fabric has a total bandwidth of 6 Tb/s in each direction, although the DAQ only uses 4 Tb/s for the RUs and 3.5 Tb/s for the BUs. Like the RUs, the BUs are also commercial computers, with the same CPUs and NICs, but with additional RAM. Each BU writes the assembled events to a local 256 GB solid-state RAM disk, which can store roughly 2 minutes of data, allowing a decoupling of the event building and event filtering so that bottlenecks at the HLT can be avoided. It then sends the data to its statically-assigned dual-CPU, multi-core FUs, via a 1-10-40 GbE network for older machines and a 10-40 GbE network for newer machines. The FUs mount the RAM disk of their BU via a network file system, and run a version of the full reconstruction software used for offline processing, to

[12]S-LINK, developed by CERN in 1995, is a specification for a FIFO-like data link that can be used to connect front-end to readout devices [23].

[13]MicroTCA (μTCA) is an open modular standard of computing architecture for high speed data flow between components.

[14]Peripheral Component Interconnect eXtended (PCI-X) is a computer bus and expansion card standard.

[15]A network interface controller (NIC) is a card that connects a computer to a network.

[16]InfiniBand is a computer-networking communication standard with very high throughput and very low latency.

[17]Fourteen Data Rate (FDR) InfiniBand provides a 14 Gb/s data rate per lane. Most InfiniBand ports are 4-lane ports allowing a speed of 56 Gb/s.

[18]A Clos network is a multistage circuit switching network that connects each input with every output with a reduced number of connections.

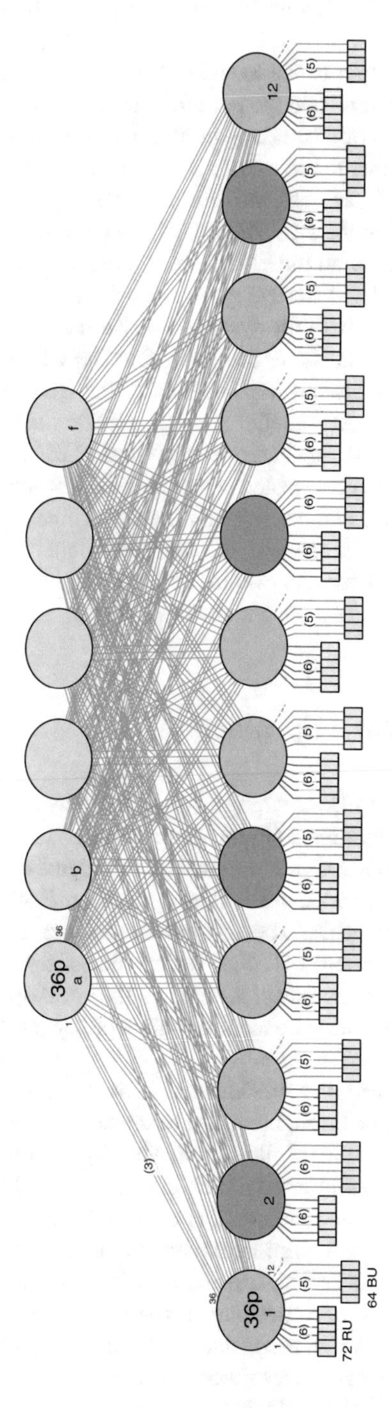

Fig. 3.12 Schematic diagram of a Clos network, connecting 72 RUs with 64 BUs [22]. Each leaf switch (1–12) uses 6 ports to connect the RUs, 6 ports for the BUs, and 3 ports to connect to a spine switch (a–f), thus allowing communication with any other RU or BU. The actual InfiniBand Clos network used in the DAQ event builder in 2016 was larger than this, with 108+108 I/O ports, connecting 108 RUs with 72 BUs

select events based on the HLT menu. The ensemble of a BU with its 8 or 16 FUs is called an appliance.

The FUs write their output to the local hard disk of the machine in the form of files, with each core producing one file per HLT stream for each luminosity section, an approximate 23 s period of data taking used as the quantum for data certification. The output files are then merged in two stages. The first stage, executed on the FU, merges the output files of each stream from all cores of the FU and copies them back to a 2 TB hard disk on the BU. The seconds stage runs on the BU and merges the per-stream output files of all the FUs in the appliance and then copies them, via the Storage Manager, to a local storage system, labelled as cluster file system (CFS) in Fig. 3.11, which runs a Lustre file system. The Storage Manager allows all BUs to simultaneously write to the same file, such that the CFS only needs to sustain a total write throughput of around 3 GB/s from all the BUs. The Storage Manager then initiates the transfer of merged files from the CFS to the CMS Tier-0 for offline processing and permanent storage at a rate of around 1 GB/s.

The Tier-0 computing centre, located at the main CERN site, carries out an immediate *prompt reconstruction* of the data using the full offline reconstruction software and exports copies of both the raw and reconstructed data to various Tier-1 centres located around the globe. Data can eventually be deleted from the Tier-0 once full copies are held at two or more independent Tier-1 sites.

3.4.2 Flow Control and Operation

The flow of data from the FEDs to storage is controlled by the central DAQ system. It begins with each FED sending a signal via the trigger throttling system (TTS) to the fast merging modules (FMMs). The possible TTS signals are: `Busy`, `Warning`, `OutOfSync`, `Error`, `Disconnected`, and `Ready`. Each FMM merges these TTS signals from several FEDs. The merged signals are then sent to the trigger control and distribution system (TCDS), which blocks the L1 GT from sending accept signals for all TTS states except `Ready`. In the usual operating case, all FEDs will be in the `Ready` state and L1 accept signals will trigger the readout of the full detector.

The entire data flow, from the subdetectors to storage, is lossless. If the central DAQ cannot sustain the data throughput, e.g. because of bandwidth limitation in the event builder, CPU limitation in the filter farm, a hardware failure, or software crash, it propagates back pressure all the way back to one or more FEDs. If the buffers in a FED become full, e.g. if too much data is coming from the detector from noise, backgrounds, or incorrect configuration, or because of back pressure from the DAQ, the FED reports a `Busy` TTS state, and the L1 trigger is throttled. This is often a temporary issue that affects one or two bunch crossings out of a hundred, leading to a deadtime of a few percent, which is considered normal in high luminosity operation. In the case of a persistent non-`Ready` state from one or more FEDs, the DAQ system is completely blocked until the problem is resolved. This often requires data taking to be stopped by the DAQ shifter, and one or more problematic subsystems to be reconfigured. In rare cases, the problem is caused by a more serious error that requires expert intervention.

The DAQ system is operated from the CMS control room, by a dedicated DAQ shifter covering 24 h per day, 7 days per week, during normal operation. The main responsibilities of the shifter are as follows:

- To initialise the DAQ system and configure all subsystems.
- To add or remove subdetectors from the central DAQ as required.
- To select the correct mode for data taking based on the current LHC status: no beam, circulating beam, or collisions.
- To start the data taking, referred to as a *run*.
- To set and adjust the rate of random triggers, which initiate a full readout of the detector irrespective of the L1 trigger decision.
- To monitor the live flow of data, from readout to event building, filtering and merging, and final transfers to the Tier-0.
- To stop the run on a change of LHC status, on request from run coordination, or in the event of an error.
- To troubleshoot errors and restart the run once solved.
- To reconfigure subsystems on request of an expert for firmware updates or troubleshooting.

I was an active DAQ shifter in both 2015 and 2016. If a DAQ error arises that cannot be solved by the shifter, a DAQ or subdetector expert-on-call is required to intervene. The DAQ expert can reprogram the DAQ and adjust internal settings, while subdetector experts can reprogram the firmware of the respective subdetector.

Overall, the DAQ is one of the best performing subsystems of CMS. It is able to manage a high throughput rate, enabling access to a large number of interesting physics events. In 2015, the CMS DAQ system processed and stored 1.2 PB of proton-proton collision data, 1.9 PB of heavy ion collision data, and 1.3 PB of auxiliary data, e.g. cosmic ray data, and detector specific information for calibration and monitoring. In 2016, this increased by around a factor of 10, with maximum read/write rate to the storage system of around 6 GB/s during data taking for the heavy-ion collisions.

References

1. Evans L, Bryant P (2008) LHC machine. JINST 3:S08001. https://doi.org/10.1088/1748-0221/3/08/S08001
2. Accelerators and Schedules. https://beams.web.cern.ch/content/accelerators-schedules. Retrieved 3 Jan 2018
3. LHC Operation in 2016. https://indico.cern.ch/event/609486/contributions/2457494/attachments/1432428/2200931/PSBLongEmitAndFutureLHCBeams.pdf. Accessed 3 Jan 2018
4. ALICE Collaboration (2008) The ALICE experiment at the CERN LHC. JINST 3:S08002. https://doi.org/10.1088/1748-0221/3/08/S08002

5. ATLAS Collaboration (2008) The ATLAS experiment at the CERN large hadron collider. JINST 3:S08003. https://doi.org/10.1088/1748-0221/3/08/S08003
6. LHCb Collaboration (2008) The LHCb detector at the LHC. JINST 3:S08005. https://doi.org/10.1088/1748-0221/3/08/S08005
7. TOTEM Collaboration (2008) The TOTEM experiment at the CERN large hadron collider. JINST 3:S08007. https://doi.org/10.1088/1748-0221/3/08/S08007
8. CMS Collaboration (2008) The CMS experiment at the CERN LHC. JINST 3:S08004. https://doi.org/10.1088/1748-0221/3/08/S08004
9. CMS Collaboration (2017) Particle-flow reconstruction and global event description with the CMS detector. JINST 12:P10003. https://doi.org/10.1088/1748-0221/12/10/P10003. arXiv:1706.04965
10. CMS Collaboration (2012) A new boson with a mass of 125 GeV observed with the CMS experiment at the large hadron collider. Science 338:1569. https://doi.org/10.1126/science.1230816
11. CMS Collaboration (2010) Alignment of the CMS silicon tracker during commissioning with cosmic rays. JINST 5:T03009. https://doi.org/10.1088/1748-0221/5/03/T03009. arXiv:0910.2505
12. CMS Collaboration (2014) Description and performance of track and primary-vertex reconstruction with the CMS tracker. JINST 9:P10009. https://doi.org/10.1088/1748-0221/9/10/P10009. arXiv:1405.6569
13. CMS Collaboration (2005) Track reconstruction in the CMS tracker. CMS-NOTE-2006-041
14. Adzic P et al (2007) Energy resolution of the barrel of the CMS electromagnetic calorimeter. JINST 2:P04004. https://doi.org/10.1088/1748-0221/2/04/P04004
15. CMS Collaboration (2008) Design, performance, and calibration of CMS hadron endcap calorimeters. CMS-NOTE-2008-010. https://cds.cern.ch/record/1103003
16. CMS, ECAL/HCAL Collaboration (2009) The CMS barrel calorimeter response to particle beams from 2-GeV/c to 350-GeV/c. Eur Phys J C 60:359. https://doi.org/10.1140/epjc/s10052-009-1024-0. [Erratum: Eur Phys J C 61:353 (2009)]
17. CMS Collaboration (2016) Performance of the CMS muon detectors in 2016 collision runs. CMS-DP-2016-046. https://cds.cern.ch/record/2202964
18. CMS Collaboration (2017) The CMS trigger system. JINST 12:P01020. https://doi.org/10.1088/1748-0221/12/01/P01020. arXiv:1609.02366
19. STEAM Trigger Menu Development. https://twiki.cern.ch/twiki/bin/view/CMS/WorkingTMDMenus, r171, https://docs.google.com/spreadsheets/d/1V1cPtokqfiHumWAg8bBwIoQn2gbJk3ka0Ppe1C27SkM. CMS Internal. Accessed 3 Jan 2018
20. CMS Trigger Modes. https://cmswbm.cern.ch/cmsdb/servlet/TriggerMode?KEY=l1_hlt_collisions2016/v414. CMS Internal. Accessed 3 Jan 2018
21. CMS Collaboration (2002) CMS: the TriDAS project. Technical design report, vol 2: data acquisition and high-level trigger. CERN-LHCC-2002-026. https://cds.cern.ch/record/578006
22. Bawej T et al (2015) The new CMS DAQ system for run-2 of the LHC. IEEE Trans Nucl Sci 62:1099. https://doi.org/10.1109/TNS.2015.2426216
23. CERN S-LINK homepage. http://hsi.web.cern.ch/HSI/s-link/. Accessed 3 Jan 2018

Chapter 4
Trigger and Object Reconstruction

For this analysis dedicated high level trigger (HLT) paths based on particle-flow jets (PFJets, see Sect. 4.3.1) have been developed. The trigger decision for these paths is made in stages, beginning with the hardware based level 1 (L1) trigger and ending with the software based HLT. I developed the HLT trigger paths myself, with initial planning and testing being made in 2014, changes for first data taking in 2015 and final adjustments throughout the 2016 data taking period.

In this chapter, a description of the L1 triggers used and the HLT paths developed for this analysis is given. The offline reconstruction of particles is then described including details of the particle flow algorithm and the final selection criteria used to define the particles. Finally, some techniques used to identify jets originating from b jets and from gluons are discussed.

4.1 Level 1 Trigger Selection

The L1 trigger must make a fast decision using only partial event information, and therefore any inference from the L1 triggers used to seed the HLT paths cannot be expected to be precise. In particular, the L1 jet reconstruction is very crude, using large initial calorimeter-energy thresholds and coarse clustering. The final state of eight jets is unlikely to be detected at the L1 trigger level, because of the low p_{T} of the non-leading jets. Nevertheless the sum of the p_{T} of all the jets in an event, H_{T}, is expected to be large as shown in Fig. 4.1, which gives the distribution of fully reconstructed H_{T} in simulated $\mathrm{t\bar{t}H}$ ($\mathrm{H} \rightarrow \mathrm{b\bar{b}}$) events using jets with $p_{\mathrm{T}} > 15\,\mathrm{GeV}$. Since the H_{T} is generally well above 200 GeV, it can often be detected by the L1 trigger.

The initial triggers implemented for the 2015 data taking run were seeded by a single L1 trigger based on the H_{T} calculated from all jet objects with a p_{T} above

D. Salerno, *The Higgs Boson Produced With Top Quarks in Fully Hadronic Signatures*, Springer Theses, https://doi.org/10.1007/978-3-030-31257-2_4

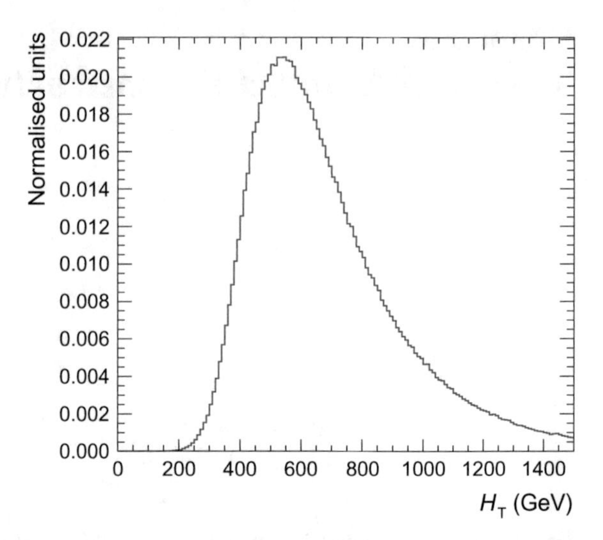

Fig. 4.1 Reconstruction-level H_T in simulated $t\bar{t}H$ ($H \rightarrow b\bar{b}$) events, using jets with $p_T > 15$ GeV and $|\eta| < 4.7$. (Repeated in Fig. 6.8.)

30 GeV and $|\eta| < 3.0$ [1]. The H_T threshold of this seed was 175 GeV, however the reconstruction of the L1 objects was quite crude at this time due to an incomplete upgrade of the L1 calorimeter trigger [2], and thus the H_T measured by the L1 trigger was significantly below the true H_T of the event, which is defined as the sum of the true p_T of all final-state jets, no matter how small.

With the 2016 data taking run the L1 calorimeter trigger upgrade was completed, which lead to a completely new L1 trigger menu. The effect of the L1 upgrade on H_T based triggers was to increase the amount of measured H_T to more accurately reflect the H_T of the event measured with the offline reconstruction. The result was that the H_T threshold for the trigger paths changed to 280 GeV. As a precaution against prescaling or disabling triggers in the face of high luminosity, a second H_T trigger with a threshold of 300 GeV has been added in a logical "OR" combination. Finally, toward the end of 2016, as the instantaneous luminosity delivered by the LHC increased, lower threshold triggers became prescaled for all trigger types, which lead to the addition of a third seed to the mix, with an H_T threshold of 320 GeV. The final L1 seed was thus a logical "OR" of three triggers with H_T thresholds of 280, 300 and 320 GeV. For part of the 2016 data taking period, the 280 GeV L1 seed was unprescaled and so the effective threshold on L1 HT was 280 GeV. Toward the end of the year the effective L1 H_T threshold was at 300 GeV at the highest instantaneous luminosities, which decreased to 280 GeV as the luminosity dropped throughout the fill. Fortunately, the 300 GeV L1 seed was never prescaled or disabled.

The rate (number of triggers per second) and efficiency (proportion of $t\bar{t}H$ events selected) of the L1 triggers was closely monitored throughout the data taking period. This is required to ensure the smooth operation of the seeds and determine if any changes need to be made to the L1 or HLT trigger paths. Figure 4.2 shows the progression of the L1 trigger post-prescale rates over time for each of the three triggers in the L1 seed. The effective H_T threshold for the L1 seed at any given time, is determined by the seed with the highest post-prescale rate.

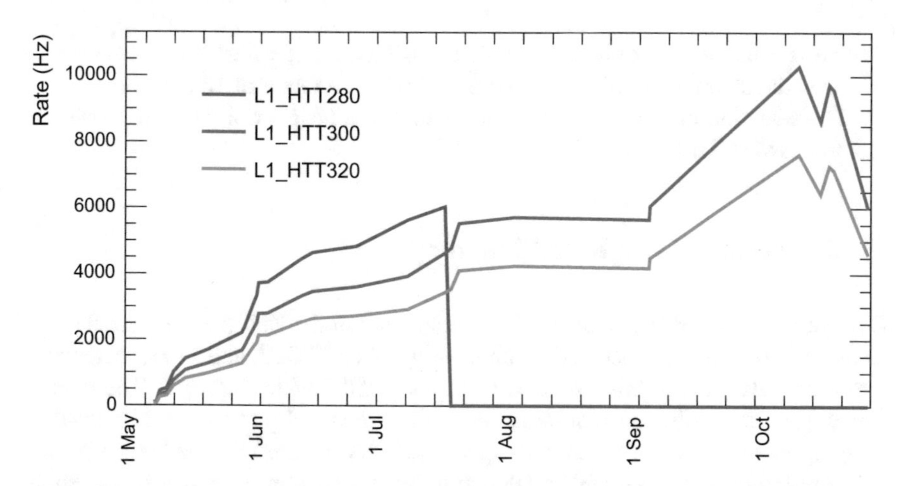

Fig. 4.2 L1 trigger post-prescale rates for the highest instantaneous luminosity over time (except for the last point). All triggers used in the L1 seed at any point of data taking are shown. The last point represents the last pp fill of the LHC in 2016

4.2 High Level Trigger Selection

After the L1 trigger selection, events are required to pass an HLT before being written to disk. In total, four HLT paths were developed for this analysis, two signal paths and two control paths, used to measure the efficiency of the signal paths. Each path processed events in two stages in order to reduce the average computation time per event. The first stage is based on quickly reconstructed calorimeter objects, while the second stage is based on the particle flow algorithm described in Sect. 4.3.1. The selection criteria of the two stages are described below. Despite the intense prescaling campaign in the face of high instantaneous luminosity as described in Sect. 4.1, the dedicated HLT signal paths used in this analysis remained unprescaled for the entire data taking period.

4.2.1 Calorimeter Based Trigger

The calorimeter based part of the HLT paths requires at least six reconstructed jets with $|\eta| < 2.6$ and p_T above 35 GeV or 25 GeV, and an H_T, calculated from jets passing this threshold, of at least 300 GeV. For the path with a lower p_T threshold, an additional requirement of a fast single b-tag jet with a combined secondary vertex (CSV) value (see Sect. 4.3.6) above 0.44 was applied. The effect of the higher threshold p_T trigger is to reduce the incoming rate to the PF component from the output rate of the L1 seed of $\sim$11 kHz to around 500 Hz—a factor of 22 reduction. The lower threshold p_T trigger reduced the incoming rate down to around 2.8 kHz, which is a factor of 4 reduction, while the addition of the fast b-tag requirement reduced the incoming rate down to around 600 Hz—a factor of 18 reduction.

The selection thresholds of the calorimeter-based trigger of each HLT path are summarised in Table 4.1.

4.2.2 Particle Flow Based Trigger

The particle flow (PF) (see Sect. 4.3.1) component of the HLT paths required six or more PFJets with $|\eta| < 2.6$ and p_T above 40 GeV or 30 GeV, and an H_T, calculated from the jets passing this first cut, of at least 450 GeV or 400 GeV. The control paths have no additional requirements, while the signal paths have additional b-tagging requirements of a single b-tagged jet (CSV > 0.63) for the higher p_T and H_T threshold path and a double b-tag requirement for the lower p_T and H_T threshold path.

The selection thresholds of each stage of each HLT path are summarised in Table 4.1. The benefit of having two signal paths with different thresholds on the selection variables is that where one path lacks efficiency, the other can compensate. For example, path 1 in the table has a low efficiency for events with a 6th jet p_T of around 40 GeV and also events with an H_T of around 450 GeV, but a high efficiency for events with 2 b jets. Conversely, path 2 will have a higher efficiency for events with a 6th jet of 40 GeV, since this is 10 GeV above the threshold, and also a higher efficiency for events with $H_T \sim 450$ GeV as this is 50 GeV above the threshold. However it has a poor efficiency for events with only 2 b-tagged jets.

In order to take advantage of the strengths of both signal paths, the analysis employs them both in an "OR" configuration. Events are then required to pass either of the two signal paths, which results in a signal efficiency of 52% for all $t\bar{t}H$ ($H \to b\bar{b}$) decays and 63% for the fully hadronic decays.

Table 4.1 Summary of the HLT paths including the thresholds used at each stage of selection. All jets considered have an $|\eta| < 2.6$

Path	Type	Calorimeter selection	Particle flow selection	Efficiency on $t\bar{t}H$ ($H \to b\bar{b}$)	
				All (%)	Hadronic (%)
1	Signal	≥6 jet, $p_T > 35$ GeV $H_T > 300$ GeV	≥6 jet, $p_T > 40$ GeV $H_T > 450$ GeV ≥1 jet, CSV > 0.63	32	42
2	Signal	≥6 jet, $p_T > 25$ GeV $H_T > 300$ GeV ≥1 jet, CSV > 0.44	≥6 jet, $p_T > 30$ GeV $H_T > 400$ GeV ≥2 jet, CSV > 0.63	49	60
3	Control	≥6 jet, $p_T > 35$ GeV $H_T > 300$ GeV	≥6 jet, $p_T > 40$ GeV $H_T > 450$ GeV	32	42
4	Control	≥6 jet, $p_T > 25$ GeV $H_T > 300$ GeV	≥6 jet, $p_T > 30$ GeV $H_T > 400$ GeV	55	67

The selection made after events are recorded is referred to as the *offline* selection, and is generally made to ensure a high efficiency of the triggers. A high trigger efficiency makes simulation modelling more robust, as there is less possibility for large differences in trigger efficiency between data and simulation. The offline selection is made to be tighter than the trigger selection and therefore drives the overall inefficiency of the search, as a portion of signal events must be cut away. In this regard, the development of the trigger involved a delicate tradeoff between signal efficiency and trigger rate, considering the necessary offline selection. The final choices of the paths are the result of negotiations with the coordinators of the trigger group and its various subgroups. When compared to the offline preselection, described in Sect. 6.3.1, the efficiency of the OR of both signal paths is 99.0%. For the final signal region discussed in Sect. 6.3.2, the trigger efficiency on the $\mathrm{t\bar{t}H}$ ($\mathrm{H} \rightarrow \mathrm{b\bar{b}}$) signal is above 99.5%.

Using the specially developed signal paths, a drop in efficiency in data at high H_{T} was observed. The drop was attributed to the last run period of the LHC in 2016 (Run H) which had a very high instantaneous luminosity and resulted in many H_{T} and missing transverse energy (MET) triggers having a rate much higher than expected. The high luminosity also caused a problem in all L1 H_{T} triggers, in which saturated (high p_{T}) jets were excluded from the H_{T} calculation. A partial mitigation strategy involves using a single jet trigger with a p_{T} threshold of 450 GeV to recover events that would have fallen short of the H_{T} threshold if such a high p_{T} jet were excluded. This strategy was adopted in this analysis and has led to the recovery of most of the lost efficiency at high H_{T}.

4.3 Object Reconstruction

The analysis requires all final-state particles produced in the $\mathrm{t\bar{t}H}$ ($\mathrm{H} \rightarrow \mathrm{b\bar{b}}$) process to be reconstructed. For the targeted signal, these particles are quarks which are reconstructed as jets, which represents the physical object. Additional information about the quarks that originate the jets is useful in the search and is estimated with the use of dedicated algorithms for b jet discrimination and quark-gluon discrimination. In order to veto events that are not compatible with the fully hadronic decay of the top-quark system, the reconstruction of leptons, specifically muons and electrons, is also necessary.

In this section the software algorithm used to reconstruct particles in the detector is explained, followed by a detailed description of the particle definitions made for this analysis. Finally, the algorithms for jet identification are described.

4.3.1 Particle Flow Algorithm

The particle flow (PF) algorithm [3] reconstructs particles passing through the detector by combining information from all layers of all subdetectors. The plethora of standard model particles can be reconstructed in the detector as a limited number of physics objects. In traditional hadron collider detectors, the subdetectors are used to measure their nominal physics objects as follows:

- Jets consisting of hadrons (mesons and baryons) and photons from the hadronisation of quarks, can be inclusively measured by the calorimeters, by considering the amount of energy deposited and the position of the calorimeter clusters.
- Isolated photons and electrons can be measured by the ECAL.
- Tagging of jets from hadronic τ decays and from b quark hadronisation is performed by the tracker, considering the tracks of the charged particles pertaining to the jet.
- Muons can be identified and measured by the muon detectors.
- Missing transverse momentum is calculated as the negative vector sum in the transverse plane of all reconstructed objects.

The PF algorithm, however, achieves a significantly improved event description by combining measurements of all subdetectors in a global reconstruction of all physics objects in the event. The reconstruction of physics objects is best explained in the context of jet reconstruction. Jets are formed by combining hadrons and potentially photons, electrons and muons which are spatially close to each other, typically within an η–ϕ cone of radius $\Delta R = \sqrt{(\Delta\eta)^2 - (\Delta\phi)^2} = 0.4$. The individual components of the jet are, in general, reconstructed as follows:

- Charged hadrons are identified by a geometrical link between one track from the tracker and one or more calorimeter clusters, together with an absence of signal in the muon detectors.
- Neutral hadrons are identified by corresponding ECAL and HCAL clusters with no linked track.
- Photons are identified by an ECAL cluster with no corresponding track or HCAL cluster.
- Electrons are identified by a track and an ECAL cluster, with a momentum-to-energy ratio compatible with 1, and no HCAL cluster.
- Muons are identified by a track in the inner tracker linked to a track in the muon detectors

Reconstructed photons, electrons and muons that are not part of a jet are considered isolated, typically within a η–ϕ cone of varying radius depending on the level of isolation required. No attempt is made to distinguish between the various species of charged and neutral hadrons. Full details of the PF algorithm and its performance in the context of the CMS detector can be found in Ref. [3], while the main features are described in the following.

Link Algorithm

The great strength of the PF algorithm lies in its ability to combine different PF elements from the various subdetectors to extract a global event description. The fundamental core of the PF reconstruction is the *link algorithm* which provides the connection between different PF elements. It tests the compatibility of two PF elements, for example a track in the tracker and an energetic ECAL cluster, resulting from a single particle. The probability that the algorithm correctly links elements from a given particle is limited by the granularity of the various subdetectors, the number of particles to resolve per unit solid angle, and the amount of detector material traversed before the calorimeters and muon detector, which can cause kinks in the trajectory and the creation of secondary particles.

Although the link algorithm can test any pair of elements in the events, the number of pairs considered are restricted in order to prevent the computing time growing as the square of the number of particles. Each element is only paired to its nearest neighbours in the η–ϕ plane. If two elements are found to be linked, the distance between the elements is calculated in order to quantify the quality of the link. *PF blocks* of elements are then produced based on direct links between elements or indirect links through common elements. The specific requirements for linking two elements depend on their type and are described in the following.

A link between a track in the central tracker and a calorimeter cluster is established if the extrapolation of the track from its last measured hit in the tracker is within the cluster area of the calorimeters. The extrapolation is performed up to three times, extending to: the two layers of the preshower; a depth corresponding to the expected maximum of a typical longitudinal electron shower profile in the ECAL; and a depth of one interaction length in the HCAL. The cluster area is defined by the union of the areas of all its cells in the η–ϕ plane for the HCAL and ECAL barrel, or the x–y plane for the ECAL endcaps and the preshower. This area can be enlarged by up to the size of a cell in each direction to account for various gaps and uncertainties. The link distance is defined as the distance in the η–ϕ plane between the extrapolated track and the cluster position. If several HCAL clusters are linked to the same track, or if an ECAL cluster is linked to several tracks, then only the link with the smallest distance is kept.

To include the energy of bremsstrahlung photos from electrons, tangents to the electron tracks are extrapolated from each tracker layer to the ECAL. A cluster from a potential radiated photon is linked to the track if an extrapolated tangent falls within the cluster area as defined above and the $\Delta\eta$ between the cluster and the extrapolated track is less than 0.05. Since bremsstrahlung photons and prompt photons have a large probability of converting to an e^+e^- pair in the tracker material, a dedicated conversion finder is used to create links between any two tracks compatible with originating from a photon conversion. If the sum of the two track momenta reproduces a photon direction that is compatible with a track tangent, these two tracks are linked to the main track.

Links between calorimeter clusters are only sought between HCAL and ECAL clusters beyond the tracker acceptance, and between ECAL and preshower clusters within the preshower acceptance. A link is established if the cluster position of the more granular calorimeter (preshower or ECAL) is within the cluster area of the less granular calorimeter (ECAL or HCAL). The link distance is defined as the distance, in the η–ϕ plane for the HCAL-ECAL link and the x–y plane for the ECAL-preshower link, between the two cluster positions. Similar to the tracker-calorimeter links, if more than one HCAL cluster is linked to a single ECAL cluster, or if multiple ECAL clusters are inked to the same preshower cluster, then only the link with the smallest distance is kept. In addition, to account for the azimuthal bending of electrons in the magnetic field, ECAL *superclusters* are formed by grouping ECAL clusters reconstructed in a small window in η and a larger window in ϕ. For the purpose of linking ECAL clusters to ECAL superclusters, at least one common ECAL cell is necessary.

Charged particles can interact with the tracker material producing new particles originating from a secondary vertex. Such displaced vertices are retained as nuclear-interaction vertices if they contain at least three tracks, at most one of which is incoming from the primary vertex and has tracker hits between the two vertices. Furthermore, the invariant mass of all outgoing tracks must be greater than 0.2 GeV. All tracks sharing a common nuclear-interaction vertex that is selected are linked together.

Extrapolated tracks from the inner tracker and segments in the muon detector are linked together if they are matched in a local x–y coordinate system, defined in a plane traverse to the beam axis, where x is the more accurately measured coordinate. The matching is made if the absolute value of Δx is less than 3 cm, or if Δx is less than 4 times its uncertainty, $\sigma_{\Delta x}$.

Within each PF block, the reconstruction sequence begins with the muon identification and reconstruction as described in Sect. 4.3.3. This is followed by the electron identification and reconstruction as described in Sect. 4.3.4 and the photon identification is performed in the same step. The last objects to be reconstructed are the jets as described in Sect. 4.3.5, which include jets from hadronic τ decays. After each stage, the PF elements used to form the objects are excluded from further consideration. Thanks to the high granularity of the CMS subdetectors, the majority of the PF blocks contain only a few elements originating from just one or two particles. The computing time necessary to process the PF event reconstruction therefore increases only linearly with the particle multiplicity.

Finally, after all PF blocks have been processed and all particles have been identified, the global event reconstruction is revisited in a post-processing step that aims to rectify misidentified objects. The majority of such cases arise from an artificially large missing transverse momentum, $p_{\mathrm{T}}^{\mathrm{miss}}$, caused by a misreconstructed high-p_{T} muon. The artificially high-p_{T} muons are often attributed to cosmic muons, a severe misconstruction of the muon momentum, or charged hadron punch through and misidentification. Additionally some muons that overlap with neutral hadrons are falsely reconstructed as charged hadrons. In each case, the particles are reclassified according to set criteria if the reclassification results in a reduction of the $p_{\mathrm{T}}^{\mathrm{miss}}$ by a factor of 2 or more.

Iterative Tracking

The most difficult objects to accurately reconstruct in the detector are jets, and thus the PF algorithm is perfectly suited to this task. In fact, the momentum resolution of charged hadrons measured in the tracker is greatly superior to that obtained from the calorimeters up to a p_{T} of several hundreds of GeV. Additionally, the tracker provides a precise measurement of the charged-particle direction at the production vertex, which is deviated by the magnetic field by the time it has propagated to the calorimeters. Since on average approximately two thirds of the jet energy is carried by charged particles, the tracker is the cornerstone of jet reconstruction.

Charged hadrons which fail to be reconstructed by the tracker would have to be solely reconstructed by the calorimeters with reduced efficiency, degraded energy resolution and biassed direction. Therefore, it is important that the tracking efficiency is as close to 100% as possible. However, it is also very important that the reconstruction of fake tracks (by incorrectly associating hits) is kept small, since these would lead to large energy excesses given their randomly distributed momentum. The seemingly incompatible requirements of high tracker efficiency and low fake rate is achieved through an iterative-tracking strategy, based on the Kalman filter (KF) [4] track reconstruction algorithm. First, tracks are seeded and reconstructed with very tight criteria, leading to a moderate tracking efficiency, but a negligibly small fake rate. Then hits which are unambiguously assigned to the tracks found in the previous stage are removed and new tracks are reconstructed with slightly looser seeding criteria. The looser seeding criteria increase the tracking efficiency, while the hit removal reduces the combinatoric matching ambiguity. This procedure is repeated until the desired efficiency and fake rate is achieved. For example, after 10 iterations, tracks originating close to the beam axis are found with an efficiency of around 99.5% for isolated muons and above 90% for charged hadrons.

4.3.2 Primary Vertex

The vertex identification is performed by the tracker, which distinguishes between primary and secondary vertices. Primary vertices are those which result from proton-proton interactions and are within a cylinder of a few millimetres around the beam axis, while secondary vertices result from decays of particles with relatively long lifetimes, such as b quarks, or particle interactions with the tracker material.

The primary vertices are separated spatially along the beam axis and are reconstructed by tracing back particle tracks to a common origin. The primary vertex with the highest quadratic sum of the p_{T} of jets clustered using only charged tracks matched to the vertex and their associated missing transverse momentum, $\sum_j p_{\mathrm{T}j}^2 + (p_{\mathrm{T}}^{\mathrm{miss}})^2$, is chosen as the primary interaction vertex and represents the hard-scattering origin of interesting physics processes. The other vertices along the beam axis are considered to come from additional minimum bias proton-proton interactions in the same bunch crossing, denoted pileup events. If a charged hadron is reconstructed in the tracker and is identified as originating from a pileup vertex, it is removed from the collection of particles used to form physics objects. This procedure is widely used in jet reconstruction and is referred to as charged hadron subtraction (CHS).

4.3.3 Muons

Muons are reconstructed at CMS using both the inner tracker and the muon spectrometer. Three different types of muons can be reconstructed depending on which of these subdetectors are used:

- **standalone muon**: Hits within each DT or CSC detector are clustered to form track segments which are then used as seeds for the pattern recognition, which gathers all DT, CSC and RPC hits along the muon trajectory. A final fit is performed and the result is referred to as a *standalone-muon track*.
- **global muon**: If a standalone-muon track is matched to a track in the inner tracker (inner track), the hits from both tracks are combined and a fit is performed, thus forming a *global-muon track*.
- **tracker muon**: Each inner track with $p_T > 0.5$ GeV and total momentum $p > 2.5$ GeV is extrapolated to the muon system. If at least one muon segment is matched to the extrapolated track, the inner track becomes a *tracker-muon track*.

For muon momenta below around 10 GeV, the tracker muon reconstruction is more efficient than the global muon as the muon is less likely to penetrate through more than one muon detector plane. Above this energy however, the global-muon reconstruction is highly efficient. Global muons and tracker muons that share an inner track are merged into a single global muon. Overall, about 99% of all muons produced within the geometrical acceptance of the muon system are reconstructed as a global muon or a tracker muon. Muons reconstructed as only standalone muons have a degraded momentum resolution and are more likely to come from cosmic muons.

The PF algorithm uses both global muons and tracker muons, but not standalone muons. Isolated global muons are defined as global muons with p_T from additional inner tracks and calorimeter energy deposits within an η–ϕ cone of radius $\Delta R = 0.3$, less than 10% of the muon p_T. This isolation criterion is sufficient to reject hadrons that may be falsely reconstructed as muons, after some of their hadron shower remnants reach the muon system (punch through).

For non-isolated global muons, the following selection criteria, corresponding to the standard CMS *tight muon selection*, are applied:

- The global muon track fit must have a normalised chi squared less than 10 ($\chi^2/\text{ndof} < 10$).
- At least one muon-chamber hit is included in the global-muon track fit.
- The tracker track must be matched to muon segments in at least two muon stations.
- The inner track must have a transverse impact parameter with respect to the primary vertex of less than 2 mm ($d_{xy} < 2$ mm).
- The inner track is required to have a longitudinal distance within 5 mm of the primary vertex ($d_z < 5$ mm).
- There must be at least one hit in the pixel detector.
- There must be hits in at least six layers of the inner tracker.

- It is required that either at least three matching track segments be found in the muon detectors, or that the calorimeter deposits associated with the track be compatible with the muon hypothesis.

These selection criteria remove the majority of punch through hadrons, cosmic muons and in flight decays of hadrons as well as guarantee a good p_{T} measurement.

The momentum of the muon is set to that measured by the inner tracker if its p_{T} is less than 200 GeV, while above this threshold the momentum is set to that calculated by the fit with the smallest χ^2 from the following fits: tracker muon only; tracker and first muon detector plane; global muon; and global excluding the muon detector planes featuring a high occupancy.

The PF elements that make up these identified muons are removed from the corresponding PF block in further processing, i.e. they are not used as building elements for other particles. However the muon identification criteria can be revisited in further PF processing. For example, the momentum and energy of charged-hadron candidates, from the tracker and calorimeters respectively, are checked for compatibility and if the track momentum is significantly larger than the calibrated sum of the linked calorimeter clusters, the muon identification is remade with somewhat looser selections on the fit quality and on the hit or segment associations.

In regard to this analysis, muons are defined for the sole purpose of vetoing events containing leptons. PF reconstructed muons as defined above are selected based on their kinematic variables p_{T} and η. Additionally, a requirement on the corrected relative muon isolation is imposed. The absolute value of the isolation variable is defined as:

$$\mathrm{Iso}^{\mu} = \sum_{\Delta R<0.4} p_{\mathrm{T}}^{\mathrm{h}^{\pm}} + \max\left(0, \sum_{\Delta R<0.4} E_{\mathrm{T}}^{\mathrm{h}^{0}} + \sum_{\Delta R<0.4} E_{\mathrm{T}}^{\gamma} - \frac{1}{2} \sum_{\Delta R<0.4} p_{\mathrm{T}}^{\mathrm{pu}}\right), \quad (4.1)$$

where $p_{\mathrm{T}}^{\mathrm{h}^{\pm}}$, $E_{\mathrm{T}}^{\mathrm{h}^{0}}$, E_{T}^{γ} and $p_{\mathrm{T}}^{\mathrm{pu}}$ are the transverse momentum or energy of particles identified by the PF algorithm. The indices denote the type of particle that are considered in the sums: charged hadrons from the primary vertex ($\mathrm{h}^{\pm}$); neutral hadrons (h^{0}); photons (γ); and charged hadrons from other vertexes (pu). The sum is performed over all particles within a cone of $\Delta R = 0.4$ around the muon. The relative isolation is defined by dividing Eq. (4.1) by the muon p_{T}.

Table 4.2 summarises the set of selection criteria applied to muons in order to veto events containing leptons. The efficiency of the muon veto is around 99.3% on the fully hadronic $\mathrm{t\bar{t}H}$ ($\mathrm{H} \rightarrow \mathrm{b\bar{b}}$) signal, increasing to over 99.6% with respect to the final selection as described in Sect. 6.3.2.

Table 4.2 Summary of muon selection requirements as used in the analysis

Variable	Requirement
p_{T}	>15 GeV
$\lvert\eta\rvert$	<2.4
$\mathrm{Iso}^{\mu}/p_{\mathrm{T}}$	<0.25

4.3.4 Electrons

Electrons are reconstructed using the information from the inner tracker and the ECAL. The ECAL is used to create electron seeds based on the energy deposited in ECAL clusters. Energetic clusters with $E_T > 4\,\text{GeV}$ are considered in an η–ϕ window around the electron direction to form a *supercluster*. The η window is kept small, while the window in ϕ is extended to account for the azimuthal bending of the electron in the magnetic field. This supercluster not only measures the energy of the bending electron, but also gathers the energy of bremsstrahlung photons radiated from it. The position and energy of the supercluster are then used to form an electron seed, with an inference of its position near the interaction point.

The tracker-based electron seeding uses the iterative tracking procedure to identify electron tracks. Non-radiating electron tracks (high p_T) can be measured as efficiently as muons, while radiating electrons produce shorter and/or lower p_T tracks, which are largely recovered by using looser requirements on the number of hits and the p_T used to form a track. All tracks resulting from the iterative tracking procedure with a p_T above 2 GeV are used as potential electron seeds.

The bremsstrahlung radiation from electrons in the tracker material is exploited to differentiate electrons from charged hadrons. When the radiated energy is small, the electron track can be reconstructed across all layers of the tracker with a well behaved χ^2 and easily propagated to the ECAL where it can be matched with the corresponding supercluster. In this case, the track forms an electron seed if the ratio of the cluster energy to the track momentum is compatible with unity. In the case of radiated photons, where this requirement fails, additional steps are carried out. Soft photon emission may lead to a successful identification of all the hits along the electron trajectory, but the track fit will generally have a large χ^2 value. On the other hand, energetic photon emission causes large changes in the electron momentum, which may lead the pattern recognition to miss some of the hits, resulting in a partial reconstruction with a small number of hits. In such cases, a preselection is made on the number of hits and the χ^2 from the tracker fit and then the selected tracks are refitted with a Gaussian-sum filter (GSF) [5]. The GSF allows for more substantial energy losses along the electron trajectory and is better suited than the KF used in iterative tracking. After this fit, a boosted decision tree (BDT) discriminator is constructed using the number of hits, the χ^2 of the GSF fit and its ratio to the KF fit, the energy lost along the GSF track, and the distance from the extrapolated track to the closest ECAL supercluster. A final requirement is made on the BDT score to select electron seeds.

The electron seeds obtained from the tracker and ECAL are merged into a single collection and used in a global fit for the electron, based on a GSF fit with more parameters than the tracker only fit, before being passed to the PF reconstruction. In a given PF block, a GSF track becomes the seed for an electron candidate if the corresponding ECAL supercluster is linked to at most two additional tracks.

To exclude the possibility of hadrons being reconstructed as electrons, the sum of the energy measured in the HCAL cells within a distance of $\Delta R < 0.15$ from the supercluster position is required to be less than 10% of the supercluster energy. To ensure the energy from radiated photons is attributed to the electron energy, all ECAL clusters in the PF block which are linked to the supercluster or to one of the GSF track tangents are associated with the electron candidate. Consequently, tracks from the inner tracker which are linked to one of these ECAL clusters are associated with the candidate if the track momentum and the energy of the linked HCAL cluster are compatible with the electron hypothesis. The tracks and ECAL clusters from photon conversions which are linked to the GSF track tangents are also associated to the candidate.

The total energy collected in the ECAL supercluster and any linked clusters is corrected to account for imperfect energy measurements based on E and η. The correction is up to 25% for low p_{T} and $|\eta| \approx 1.5$, where the tracker thickness is largest. The final energy assigned to an electron is determined from combining the corrected ECAL energy and the GSF track momentum, while the direction is set to that of the GSF track.

Electron candidates are required to satisfy an additional identification requirement, which can include the use of up to 14 variables. The variables, calculated from the GSF track, ECAL and HCAL clusters, are combined in a BDT, trained separately in the ECAL barrel and endcaps and for isolated and non isolated electrons, and then a selection on the BDT score is made.

The tracks and clusters in the PF block used to reconstruct electrons are excluded from further processing. Tracks identified as originating from photon conversion which are not used in the electron reconstruction are also excluded from further processing since they are often poorly measured and are likely to be misreconstructed tracks.

As is the case for muons, in this analysis, electrons are defined for the sole purpose of vetoing events containing leptons. PF reconstructed electrons are selected based on their p_{T} and η as well as a number of isolation variables. The selection adopted in this analysis is summarised in Table 4.3 and corresponds to a rather loose electron definition. The variables listed in the table are defined as follows:

- $5 \times 5\ \sigma_{i\eta i\eta}$ is the weighted cluster RMS η along η and inside the 5×5 crystal region around the seed crystal.
- $\Delta\eta_{\mathrm{seed}}$ is the difference between the η of the track and that of the track seed, plus the difference between the η of the supercluster and that of the supercluster seed.
- $\Delta\phi_{\mathrm{in}}$ is the difference between the ϕ of the track and that of the track seed.
- H/E is the fraction of HCAL energy to ECAL energy.
- $\mathrm{Iso}^{e}/p_{\mathrm{T}}$ is the relative combined PF isolation with effective area correction. Its absolute value is defined as:

$$\mathrm{Iso}^{e} = \sum_{\Delta R<0.4} p_{\mathrm{T}}^{\mathrm{h}^{\pm}} + \max\left(0, \sum_{\Delta R<0.4} E_{\mathrm{T}}^{\mathrm{h}^{0}} + \sum_{\Delta R<0.4} E_{\mathrm{T}}^{\gamma} - \rho \cdot \mathrm{e}A(\eta)\right), \quad (4.2)$$

where $p_{\mathrm{T}}^{\mathrm{h}^{\pm}}$, $E_{\mathrm{T}}^{\mathrm{h}^{0}}$ and E_{T}^{γ} are the same PF variables as in Eq. (4.1), $\mathrm{e}A$ is the effective area in η–ϕ space of the ECAL clusters at a given η, and ρ is the median jet energy

Table 4.3 Summary of electron selection requirements as used in the analysis

Variable	Selection requirement	
	Barrel	Endcap
p_{T}	>15 GeV	
$\|\eta\|$	<2.4	
$\|\eta_{\mathrm{supercluster}}\|$	≤1.479	>1.479
$5 \times 5\ \sigma_{\mathrm{i}\eta\mathrm{i}\eta}$	<0.0115	<0.037
$\Delta\eta_{\mathrm{seed}}$	<0.00749	<0.00895
$\Delta\phi_{\mathrm{in}}$	<0.228	<0.213
H/E	<0.356	<0.211
$\mathrm{Iso}^{e}/p_{\mathrm{T}}$	<0.175	<0.159
$\|1/E - 1/p\|$	<0.299	<0.15
d_{xy}	<0.05	<0.10
d_z	<0.10	<0.20
$N_{\mathrm{missing\_hits}}$	≤2	≤3
`Pass conversion veto`	✓	✓

per unit area of all jets in the event. The effective area correction corrects for pileup effects.

- $|1/E - 1/p|$ is the absolute difference between the inverse of the ECAL energy and the inverse of the track momentum.
- d_{xy} is the impact parameter distance from the primary vertex in the x–y plane.
- d_z is the impact parameter distance form the primary vertex along the beam axis.
- $N_{\mathrm{missing\_hits}}$ is the number of missing hits in the tracker expected given the electron hypothesis.
- `Pass conversion veto` is true if there are no photon conversion vertices associated with the cluster, i.e. the electron is not from a photon conversion.

The efficiency of the electron veto is around 99.2% on the fully hadronic $\mathrm{t\bar{t}H}$ ($\mathrm{H} \rightarrow \mathrm{b\bar{b}}$) signal, increasing to over 99.5% with respect to the final selection as described in Sect. 6.3.2.

4.3.5 Jets

Jets arise from the hadronisation of quarks and consist of many closely spaced particles which are detected as charged hadrons (e.g. $\pi^{\pm}$, $\mathrm{K}^{\pm}$, or protons), neutral hadrons (e.g. $\mathrm{K}^0_{\mathrm{L}}$ or neutrons), non isolated photons (e.g. from π^0 decays), and less often additional muons or electrons from decays of charged hadrons.

Within the tracker acceptance ($|\eta| < 2.5$), all ECAL clusters not linked to a track are considered photons, while all HCAL clusters without a linked track are considered neutral hadrons. The assumption that neutral ECAL energy deposits are from photons is justified since photons carry 25% of the energy of hadronic jets, while neutral hadrons deposit only 3% of the jet energy in the ECAL. However, beyond the tracker acceptance, charged hadrons cannot be distinguished from neutral hadrons and they deposit over 20% of the jet energy in the ECAL. For this reason, ECAL clusters linked to an HCAL cluster are considered to arise from the same hadron shower, without distinction between charged or neutral hadrons, while photons are attributed to the ECAL clusters without an HCAL link. HCAL clusters without an ECAL link are naturally considered hadrons. The energies measured in the ECAL and HCAL are calibrated as described in Sects. 3.2.2 and 3.2.3 to provide the final neutral hadron and photon energies. In the forward region, the HF EM and HF HAD clusters are considered *HF photons* and *HF hadrons* respectively, without any calibration.

The remaining HCAL clusters in the PF block, necessarily within the tracker acceptance, are linked to one or more tracks,[1] which can in turn be linked to one or more ECAL clusters.[2] The measured HCAL and ECAL energies are calibrated as described in Sect. 3.2.3 to determine the calibrated calorimeter energy, which is then compared to the sum of the momenta of the linked tracks to determined the particle content and final jet energy, as described in the following.

If the calibrated calorimeter energy is greater than the sum of the track momenta by more than the calorimetric energy resolution for hadrons, the excess is attributed to photons or neutral hadrons. If the excess is between 500 MeV and the total ECAL energy, it is attributed to a photon with an energy equal to the recalibrated excess under the photon hypothesis described in Sect. 3.2.2. If the excess is larger than the total ECAL energy, the recalibrated ECAL energy is attributed to a photon and the difference, if more than 1 GeV, is attributed to a neutral hadron. Each track is attributed to a charged hadron, with a momentum and energy derived from the track momentum assuming a charged-pion mass.

If the calibrated calorimeter energy, within its resolution, is compatible with the sum of the track momenta, no neutral particle is identified. The momenta of the charged hadrons are recalculated based on a χ^2 fit of the tracker momenta and the calorimeter energies. The fit reduces to a weighted average if only one track is linked to the HCAL cluster. The combination of subdetector measurements is particularly important when the track momentum is poorly measured, e.g. at high energy or large η, and ensures a smooth energy resolution across the low-energy and high-energy regimes, dominated by the tracker and calorimeter measurements respectively. Even at the highest energies, the energy resolution resulting from the combined calculation is superior to that from the calorimeter alone.

In some rare cases, it may be that the calibrated calorimeter energy is less than three standard deviations below the sum of the track momenta. In such cases, a second search for muons with relaxed identification criteria is performed and all resulting global muons are considered PF muons, if the relative precision on their momentum is less than 25%, and the corresponding tracks are removed from the jet. This allows

[1]Each of these tracks is necessarily not linked to another HCAL cluster.

[2]Each of these ECAL clusters is necessarily linked to only one track.

a few more muons to be found without increasing the misidentification rate, since the calorimeter energy is too low for punch-through hadrons to be misreconstructed as muons. If the sum of the momenta of the remaining jet tracks is still significantly larger than the calibrated calorimeter energy, the excess is likely to arise from misreconstructed tracks with a p_{T} uncertainty greater than 1 GeV. These tracks are ranked according to their p_{T} uncertainty and sequentially removed until the momentum excess vanishes or no such tracks remain, whichever comes first. Less than 0.03% of the tracks in simulated QCD multijet events are affected by this track removal procedure. In general, after the muon and track removal, the calibrated calorimeter energy is either compatible with the reduced sum of track momenta, or larger than it. The two cases are then treated as described above.

The PF reconstructed jet constituents described above are clustered together to form jets using the anti-k_{T} algorithm with a radius parameter of 0.4 (AK4). The algorithm is described in detail in Ref. [6], while a brief overview is given here. The clustering begins by defining distance parameters d_{ij} between entities (individual or combinations of particles) i and j and $d_{i\mathrm{B}}$ between entity i and the beam axis (B):

$$d_{ij} = \min(p_{\mathrm{T}i}^{-2}, p_{\mathrm{T}j}^{-2})\frac{\Delta_{ij}^2}{R^2}, \quad (4.3)$$

$$d_{i\mathrm{B}} = p_{\mathrm{T}i}^{-2} \quad (4.4)$$

where $\Delta_{ij}^2 = (y_i - y_j)^2 + (\phi_i - \phi_j)^2$, R is the user defined radius parameter, and $p_{\mathrm{T}i}$, y_i and ϕ_i are the transverse momentum, rapidity and azimuth, respectively, of entity i.

The clustering proceeds by calculating the distances between all particle pairs and between each particle and the beam axis. If the smallest distance is d_{ij}, then particles i and j are combined to form a single entity. On the other hand, if the smallest distance is $d_{i\mathrm{B}}$, then entity i is considered a jet and removed from the list of entities. After each stage the distances are recalculated and the procedure is repeated until no entities remain. This algorithm typically results in particles with high p_{T} accumulating all low-p_{T} particles within a cone of radius R. In the event of overlapping high-p_{T} particles ($\Delta_{12} < R$), either one particle will dominate forming a conical jet or the jet will be a union of three cones: two with radius less than R centred on each particle and one with radius R centred on the final jet. In event of two close high-p_{T} particles ($R < \Delta_{12} < 2R$), either one particle will dominate forming a conical jet while the other forms a conical jet with the overlapping part missing, or neither jet will be conical as they share the overlapping part.

The anti-k_{T} algorithm outperforms similar algorithms (k_{T} and Cambridge/Aachen) and iterative-cone algorithms [6] and is thus used as the standard jet clustering algorithm in CMS. It is infrared and collinear safe,[3] its jet boundaries are resilient to soft radiation, and its jet momenta are resilient to smearing caused by pileup.

[3]Infrared safe means that no infrared singularities appear in the perturbative calculations and that jet solutions are insensitive to soft radiation. Collinear safe means that no collinear singularities appear in the perturbative calculations and that solutions are insensitive to collinear radiation.

In CMS CHS is employed before initiating the jet clustering. The resulting jets are then subjected to the following selection criteria, corresponding to the standard CMS *loose jet identification*, to increase the purity of the jet:

- Absolute value of η less than 2.4
- Fraction of energy attributed to neutral hadrons less than 0.99
- Fraction of energy attributed to photons less than 0.99
- Fraction of energy attributed to electrons less than 0.99
- Non-zero fraction of energy attributed to charged hadrons
- At least two constituents
- At least one charged constituent.

After these selections are made, the jet energies are calibrated in terms of the jet energy scale (JES) and the jet energy resolution (JER). The details of the calibration are described in Ref. [7] and the main features are summarised here.

JES corrections aim to bring the measured jet energy closer to the true jet energy. They are implemented in stages and are applied to both Monte Carlo (MC) simulated events and data. The first stage is a pileup offset correction which aims to reverse the impact of pileup on the jet energy. Next comes a jet response correction which accounts for the imperfect measurement of the detector. The next stage is only applied to jets in data and corrects for residual differences between data and simulation, since the corrections are primarily derived from MC simulation. Finally, a jet-flavour correction is applied to account for differences in the quark-gluon composition of jets in the different MC samples used to derive the corrections. The derivation of the total JES correction is subject to a number of systematic uncertainties which are propagated to the event selection and considered in the final result. The treatment of the JES uncertainties is described, along with all other uncertainties affecting the analysis, in Sect. 6.6.

The JER observed in data is worse than that predicted by MC simulation and thus jets in simulation are smeared to better reflect the data. The JER correction for each jet is calculated as a function of the p_T and η of the jet and the p_T of the clustered generator-level particles, if they are matched to the jet particles. If there is no generator level particles matched to the jet, a stochastic smearing is applied based on the η of the jet and the resolution of its p_T, which aims to reproduce the average data-to-simulation difference. The uncertainty in the derivation of the JER is also considered in the event selection and final result.

After the jets have been identified, reconstructed, calibrated and corrected, a final selection is made on their p_T and η. In this analysis, jets are considered if they have $p_T > 30\,\text{GeV}$ and $|\eta| < 2.4$.

4.3.6 B-Tagging

Jets originating from the hadronisation of b quarks are able to be identified with dedicated *b-tagging algorithms*, which exploit the relatively long lifetime of b quarks.

A number of different tagging algorithms are employed at CMS, differing in their performance, robustness, and validation. The choice for this analysis is the combined secondary vertex (CSV) algorithm [8, 9], which combines information on the primary vertex, impact parameters, and secondary vertices within the jet using a neural network discriminator. A schematic diagram of these parameters is given in Fig. 4.3.

Many different variables are used in the input to the neural network, such as the track multiplicity, secondary vertex multiplicity, the fraction of energy carried by tracks at the vertex, the impact parameter significance of tracks, and the corrected mass of the secondary vertex, defined as $\sqrt{M_{\mathrm{SV}}^2 + p^2 \sin^2\theta} + p\sin\theta$, where M_{SV} is the invariant mass of the tracks from the secondary vertex with momentum p, and θ is the angle between p and the vector connecting the primary and secondary vertices. The latter two variables are shown in Fig. 4.4 for jets with p_{T} above 20 GeV in simulated $\mathrm{t\bar{t}}$ events.

The output of the CSV algorithm is a discriminator for each jet, which lies between zero and one. A value of the CSV discriminator close to 1 indicates a high probability that the jet originates from a bottom quark, while a value close to 0 indicates a

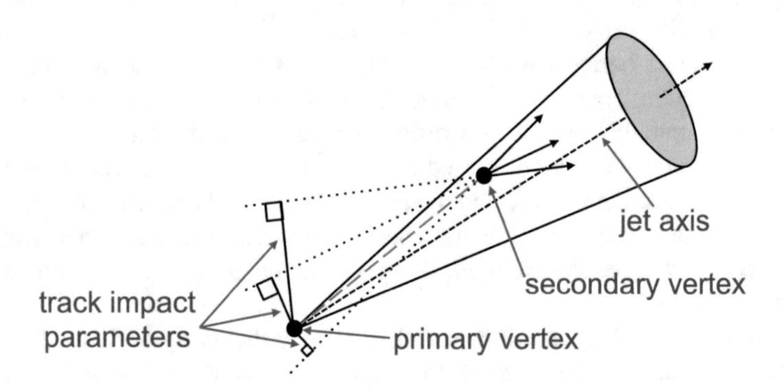

Fig. 4.3 Schematic diagram of a jet originating from a bottom quark, showing the definitions of the primary vertex, the secondary vertex, and the track impact parameters

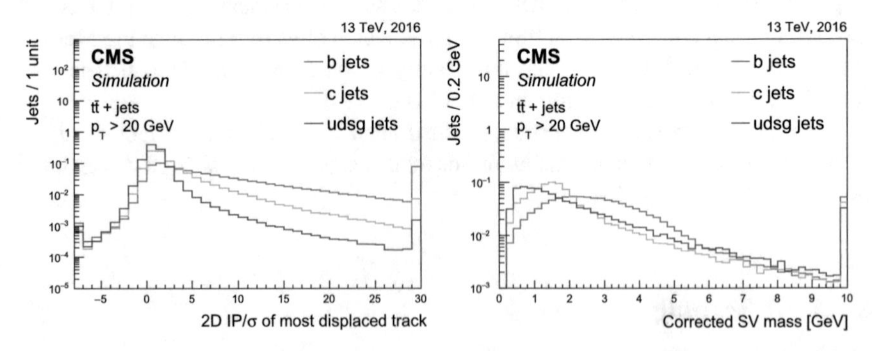

(a) Impact parameter significance of first track (b) Corrected mass of the secondary vertex

Fig. 4.4 Selected CSV input variables for jets with $p_{\mathrm{T}} > 20\,\mathrm{GeV}$ in simulated $\mathrm{t\bar{t}}$ events [9]

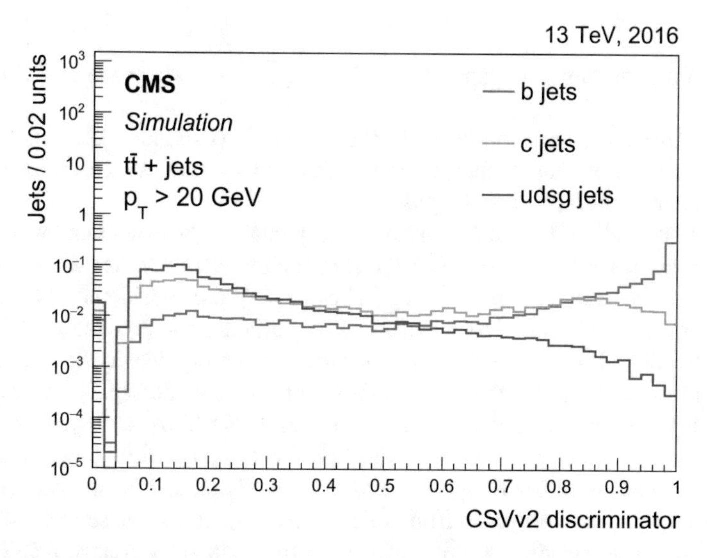

Fig. 4.5 CSV discriminator output for jets with $p_{\mathrm{T}} > 20\,\mathrm{GeV}$ in simulated $\mathrm{t\bar{t}}$ events [9]

large probability that the jet originates from a light-flavour quark or gluon. This is displayed in Fig. 4.5, which shows the CSV discriminator for jets with p_{T} above 20 GeV in simulated $\mathrm{t\bar{t}}$ events.

Two different working points of the CSV discriminator are used in this analysis. In simulated $\mathrm{t\bar{t}}$ events, the loose (medium) working point has an efficiency of around 81% (63%) to tag jets with $p_{\mathrm{T}} > 20\,\mathrm{GeV}$ originating from b quarks, 37% (12%) for jets originating from c quarks, and approximately 9% (1%) misidentification probability for jets from light-flavour quarks or gluons. In the remainder of this document, a jet passing the loose (medium) b-tagging working point is referred to as CSVL (CSVM).

4.3.7 Quark-Gluon Discrimination

A further classification of jets can be made that aims to identify those originating from the hadronisation of gluons. Specifically, CMS has developed an algorithm [10, 11] that can discriminate between jets originating from gluons and those originating from light-flavour quarks, i.e. u, d and s quarks. The algorithm, denoted *quark-gluon likelihood* (QGL), is a likelihood discriminant formed from three variables based on the PF jet constituents:

- Multiplicity: the total number of PF candidates reconstructed within the jet.
- Axis2: the RMS minor axis in the η–ϕ plane of the PF candidates (i.e. the minor axis of the jet ellipse).

- Fragmentation function: defined as $p_{\mathrm{T}}D = \frac{\sqrt{\sum_i p_{\mathrm{T},i}^2}}{\sum_i p_{\mathrm{T},i}}$, where i is a PF candidate.

The choice of variables is motivated by the observation that gluon jets typically have a higher number of constituents, are less collimated and have a softer fragmentation function than light-flavour-quark jets.

When computing the above variables, only charged hadrons that are linked to a high-purity track compatible with the primary interaction vertex are considered. Similarly, only neutral hadrons with p_{T} above 1 GeV are considered. At least three PF jet particles are required to calculate all the variables and thus the likelihood.

The likelihood itself is the product of three probability density functions (pdfs) of each of these variables. The pdfs are derived from simulated QCD dijet events, in which the jet origins are and are computed in bins of η and p_{T}. Only jets originating from u, d and s quarks are used for the light-flavour jets, while those originating from gluons are used for the gluon jets. The likelihood can be interpreted as the probability of a jet to originate from a light-flavour quark, so these jets will have a QGL distribution peaking at unity, while jets from gluons will have a QGL value peaking at zero. The distributions of the input variables and the QGL output for simulated QCD dijet events used to derive the pdfs are shown in Fig. 4.6.

The QGL is used in this analysis to separate events containing light-flavour jets from the hadronic decay of the W boson from events containing gluon jets produced in QCD interactions. Specifically, an event-based, rather than jet-based, likelihood ratio discriminant is formed, using the QGL information for jets which are not identified as b jets. The quark-gluon likelihood ratio (QGLR) is defined as follows:

$$q_{LR(N_1 \mathrm{v} N_2)} = \frac{L(N_1, 0)}{L(N_1, 0) + L(N_2, N_1 - N_2)}, \tag{4.5}$$

where N_1 is the number of jets and also the number of quarks in the first hypothesis, and N_2 is the number of quarks in the second hypothesis. The individual likelihoods are defined as:

$$L(N_{\mathrm{q}}, N_{\mathrm{g}}) = \sum_{\mathrm{perm.}} \left(\prod_{k=i_1}^{i_{N_q}} f_{\mathrm{q}}(\zeta_k) \prod_{m=i_{N_{\mathrm{q}}+1}}^{i_{N_{\mathrm{q}}+N_{\mathrm{g}}}} f_{\mathrm{g}}(\zeta_m) \right), \tag{4.6}$$

where ζ_i is the QGL discriminator for the ith jet, and $f_{\mathrm{q(g)}}$ is the probability density function of ζ_i when the ith jet originates from a quark (gluon). The former include u, d, s, and c quarks, but not b quarks. The sum in Eq. (4.6) runs over all inequivalent permutations of assigning N_{q} jets to quarks and N_{g} jets to gluons. In this analysis the likelihood ratio $q_{LR(N\mathrm{v}0)}$ is used, which compares the likelihood of N reconstructed light jets originating from N quarks to the likelihood of N reconstructed light jets coming from N gluons. A requirement of $q_{LR(N\mathrm{v}0)} > 0.5$ is applied to events with $N = 3, 4, 5$ light jets, excluding either the first 3 or 4 b-tagged jets (by CSV value).

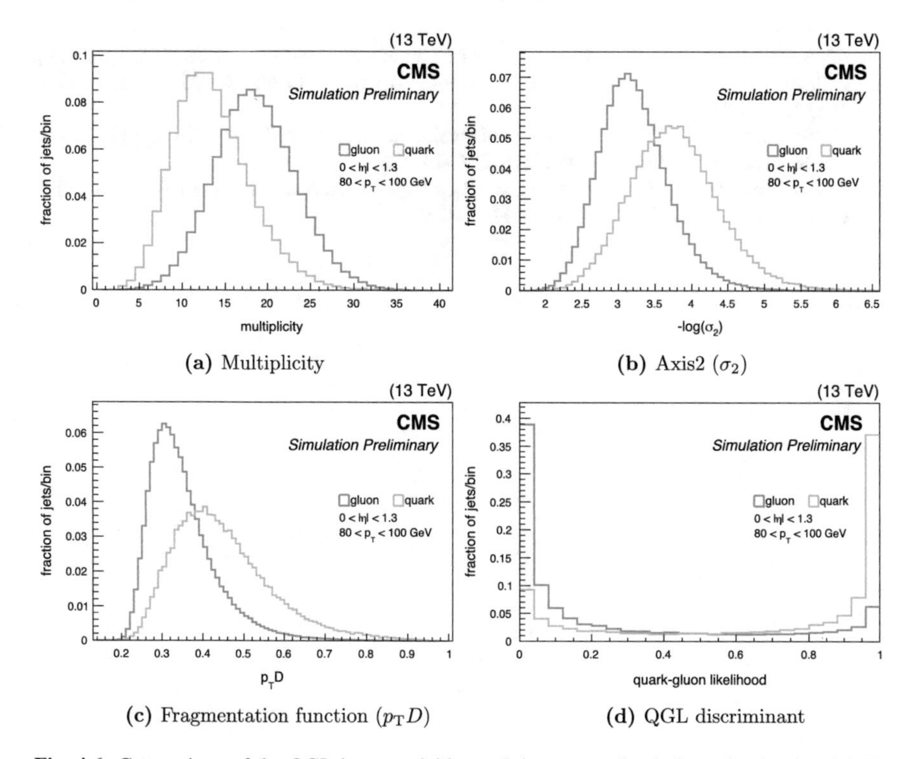

(**a**) Multiplicity (**b**) Axis2 (σ_2)

(**c**) Fragmentation function ($p_T D$) (**d**) QGL discriminant

Fig. 4.6 Comparison of the QGL input variables and the output discriminant in simulated QCD dijet events, for jets with $|\eta| < 1.3$ and $80 < p_T < 100$ GeV [11]. Expected distributions for light-flavour quark jets and gluon jets are shown separately, normalised to unity

References

1. CMS Collaboration (2016) CMS L1 calorimeter trigger performance in 2016 data. https://cds.cern.ch/record/2202966. CMS-DP-2016-044
2. CMS Collaboration (2017) The CMS level-1 trigger system for LHC Run II. JINST 12:C03021. https://doi.org/10.1088/1748-0221/12/03/C03021
3. CMS Collaboration (2017) Particle-flow reconstruction and global event description with the CMS detector. JINST 12:P10003. https://doi.org/10.1088/1748-0221/12/10/P10003, arXiv:1706.04965
4. CMS Collaboration (2005) Track reconstruction in the CMS tracker. https://cds.cern.ch/record/934067. CMS-NOTE-2006-041
5. Adam W, Frühwirth R, Strandlie A, Todor T (2005) Reconstruction of electrons with the Gaussian-Sum filter in the CMS tracker at the LHC. https://cds.cern.ch/record/815410. CMS-NOTE-2005-001
6. Cacciari M, Salam GP, Soyez G (2008) The anti-k_t jet clustering algorithm. JHEP 04:063. https://doi.org/10.1088/1126-6708/2008/04/063, arXiv:0802.1189
7. CMS Collaboration (2017) Jet energy scale and resolution in the CMS experiment in pp collisions at 8 TeV. JINST 12:P02014. https://doi.org/10.1088/1748-0221/12/02/P02014, arXiv:1607.03663
8. CMS Collaboration (2013) Identification of b-quark jets with the CMS experiment. JINST 8:P04013. https://doi.org/10.1088/1748-0221/8/04/P04013, arXiv:1211.4462

9. CMS Collaboration (2018) Identification of heavy-flavour jets with the CMS detector in pp collisions at 13 TeV. JINST 13(05):P05011. https://doi.org/10.1088/1748-0221/13/05/P05011, arXiv:1712.07158
10. CMS Collaboration (2013) Performance of quark/gluon discrimination in 8 TeV pp data. https://cds.cern.ch/record/1599732. CMS-PAS-JME-13-002
11. CMS Collaboration (2016) Performance of quark/gluon discrimination in 13 TeV data. http://cds.cern.ch/record/2234117. CMS-DP-2016-070

Chapter 5
The Matrix Element Method

The matrix element method (MEM) is a likelihood technique based on the leading order (LO) matrix element for the signal process and possibly one or more background processes. It was initially used by the D0 Collaboration for measuring the mass of the top quark [1] and the helicity of the W boson [2]. It gained prominence at the Tevatron experiments and was used in several measurements of the top quark, before being first used as a discriminant in a search-based analysis at CDF in the search for the Higgs boson [3]. The method was then successfully brought to CMS and implemented in a search for $t\bar{t}H$ production with $H \rightarrow b\bar{b}$ in the leptonic top-quark decay channels at 8 TeV [4].

The MEM uses the full kinematic properties of the event, i.e. the energy and direction of leptons and jets as well as the missing transverse momentum, to calculate the probability that an event arises from the signal process. The probability density is calculated based on the cross section formula for the processes, which includes the LO matrix element amplitude, the phase space and the parton luminosity, and a transfer function that accounts for detector effects, i.e. differences in measured and true quantities. It may also calculate a similar probability density for one or more background processes, and combine them all in a likelihood ratio.

The algorithm used in this analysis uses the matrix element of the LO gluon-initiated $t\bar{t}H$ process, described in Sect. 2.5 (Fig. 2.14b, c, and d), for the signal hypothesis and the LO gluon-initiated $t\bar{t} + b\bar{b}$ process (Fig. 2.15d) for the background hypothesis.[1] It also incorporates the hadronic decays of the Higgs boson and the top quarks to their final-state jets, giving a quark-level cross section similar to that in Eq. (2.86). Given the uncertainty of matching the detected jets with the underlying quarks, all possible jet-quark associations are considered. The total number of

[1]The simplifying assumption of initiation through gluon-gluon fusion is valid as this represents most of the yield, e.g. at $\sqrt{s} = 14$ TeV, the fraction of the gluon-initiated $t\bar{t}H$ subprocesses is about 80% of the inclusive next-to-leading-order (NLO) cross section [5].

D. Salerno, *The Higgs Boson Produced With Top Quarks in Fully Hadronic Signatures*, Springer Theses, https://doi.org/10.1007/978-3-030-31257-2_5

associations is constrained by considering only the three or four jets that are most likely to originate from b jets (according to their CSV discriminant value) explicitly as candidates for the b quarks, while the other jets are considered candidates for the light-flavour quarks. Finally, the effects of hadronisation and imperfect detector measurement are taken into account via transfer functions, which are obtained from simulated events.

The various components mentioned above are combined to calculate the following probability density function under the signal (S) or background (B) hypothesis, $i =$ S, B:

$$w_i(\vec{y}) = \int \frac{1}{s} \frac{g(x_a; Q_i) g(x_b; Q_i)}{x_a x_b} \delta^4\Big(x_a P_a + x_b P_b - \sum_{q=1}^{8} p_q\Big) |\mathcal{M}_i(\vec{x})|^2 W(\vec{y}|\vec{x}) \mathrm{d}\Phi \mathrm{d}x_a \mathrm{d}x_b, \quad (5.1)$$

where $\sqrt{s}$ is the centre-of-mass pp collision energy, $g(x; Q)$ is the gluon parton density function (PDF) evaluated at the factorisation scale Q, x_a and x_b are the momentum fractions of the initiating partons (gluons), P_a and P_b are the 4-momenta of the colliding protons (all described in Sect. 5.4), $\mathcal{M}_i(\vec{x})$ is the production and decay amplitude (defined in Sect. 5.2), $W(\vec{y}|\vec{x})$ is the transfer function (described in Sect. 5.3) and $\mathrm{d}\Phi$ is the phase space (defined in Sect. 5.1).

The probability densities for the $\mathrm{t\bar{t}H}$ and $\mathrm{t\bar{t}+b\bar{b}}$ hypotheses are combined in a likelihood ratio to form a final discriminant, which yields values closer to one for signal events and values closer to zero for background events. A detailed description of the various components leading to the construction of the final discriminant is provided in the following.

5.1 Construction of the Phase Space

The phase space of the final-state particles includes contributions from eight quarks: three from each top-quark decay and two from the Higgs-boson decay. For each top quark, t_i, with $i = 1, 2$, the phase-space volume of its decay products, with 4-momentum $\{q_i, q_i', b_i\}$, is written as:

$$\mathrm{d}\Phi_{t_i} = \frac{1}{(2\pi)^9} \frac{\mathrm{d}\vec{q}_i}{2E_{q_i}} \frac{\mathrm{d}\vec{q}_i'}{2E_{q_i'}} \frac{\mathrm{d}\vec{b}_i}{2E_{b_i}}, \quad (5.2)$$

while for the decay products of the Higgs boson, or the two additional b quarks in the case of the $\mathrm{t\bar{t}+b\bar{b}}$ hypothesis, with 4-momentum $\{b, \bar{b}\}$, the phase-space volume is written as:

$$\mathrm{d}\Phi_{b\bar{b}} = \frac{1}{(2\pi)^6} \frac{\mathrm{d}\vec{b}}{2E_b} \frac{\mathrm{d}\vec{\bar{b}}}{2E_{\bar{b}}}. \quad (5.3)$$

These expressions can be simplified upon change of coordinates by using the following relation:

$$\frac{\mathrm{d}\vec{p}}{E} = \frac{|\vec{p}|^2 \mathrm{d}|\vec{p}|\mathrm{d}\Omega}{E} = \frac{|\vec{p}| E \mathrm{d}E \mathrm{d}\Omega}{E} = |\vec{p}|\mathrm{d}E\mathrm{d}\Omega \tag{5.4}$$

where $\mathrm{d}\Omega = \mathrm{d}\cos\theta\mathrm{d}\phi$, with θ and ϕ defined in Sect. 3.2, and $|\vec{p}|\mathrm{d}|\vec{p}| = E\mathrm{d}E$ follows from $E^2 - |\vec{p}|^2 = m^2$. Equations (5.2) and (5.3) can now be rewritten as:

$$\mathrm{d}\Phi_{t_i} = \frac{1}{8(2\pi)^9} |\vec{q}_i||\vec{q}_i^{\,\prime}||\vec{b}_i| \mathrm{d}E_{q_i} \mathrm{d}E_{q_i^{\prime}} \mathrm{d}E_{b_i} \mathrm{d}\Omega_{q_i} \mathrm{d}\Omega_{q_i^{\prime}} \mathrm{d}\Omega_{b_i} \tag{5.5}$$

$$\mathrm{d}\Phi_{b\bar{b}} = \frac{1}{4(2\pi)^6} |\vec{b}||\vec{\bar{b}}| \mathrm{d}E_b \mathrm{d}E_{\bar{b}} \mathrm{d}\Omega_b \mathrm{d}\Omega_{\bar{b}}, \tag{5.6}$$

and the total phase-space volume as:

$$\mathrm{d}\Phi = \mathrm{d}\Phi_{t_1} \mathrm{d}\Phi_{t_2} \mathrm{d}\Phi_{b\bar{b}}. \tag{5.7}$$

5.1.1 *Reduction of the Dimensionality*

The 24-dimensional phase-space volume in Eq. (5.7), when coupled with the conservation of 4-momentum delta function, can be reduced in dimensionality by taking advantage of the over-constrained parameter space. In particular, the energies of all but one of the decay products of each top quark and the Higgs boson can be eliminated by replacing their energies with the invariant mass squared of particle pairs. The relevant invariant masses are defined in terms of the 4-momenta as follows:

$$\begin{aligned} M_{W_i}^2 &= m_{q_i q_i^{\prime}}^2 &&= (q_i + q_i^{\prime})^2, \\ M_{t_i}^2 &= m_{q_i q_i^{\prime} b_i}^2 &&= (q_i + q_i^{\prime} + b_i)^2, \\ M_H^2 &= m_{b\bar{b}}^2 &&= (b + \bar{b})^2, \end{aligned} \tag{5.8}$$

with $i = 1, 2$. The following change of variables can then be made:

$$(E_q, \Omega_q, E_{q^{\prime}}, \Omega_{q^{\prime}}, E_b, \Omega_b)_i \rightarrow (E_q, \Omega_q, m_{qq^{\prime}}^2, \Omega_{q^{\prime}}, m_{qq^{\prime}b}^2, \Omega_b)_i, \tag{5.9}$$

$$(E_b, \Omega_b, E_{\bar{b}}, \Omega_{\bar{b}}) \rightarrow (E_b, \Omega_b, m_{b\bar{b}}^2, \Omega_{\bar{b}}). \tag{5.10}$$

Equation (5.9) requires multiplication by the inverse of the determinant of the Jacobian matrix given by:

$$J_t = \begin{vmatrix} \frac{\partial m_{qq^{\prime}}^2}{\partial E_{q^{\prime}}} & \frac{\partial m_{qq^{\prime}b}^2}{\partial E_{q^{\prime}}} \\ \frac{\partial m_{qq^{\prime}}^2}{\partial E_{\bar{b}}} & \frac{\partial m_{qq^{\prime}b}^2}{\partial E_{\bar{b}}} \end{vmatrix}. \tag{5.11}$$

where the individual elements are evaluated from Eq. (5.8) to be:

$$
\begin{aligned}
\frac{\partial m^2_{qq'}}{\partial E_{q'}} &= \frac{m^2_{qq'}}{E_{q'}} \\
\frac{\partial m^2_{qq'}}{\partial E_b} &= 0 \\
\frac{\partial m^2_{qq'b}}{\partial E_b} &= 2(q+q')\frac{\partial b}{\partial E_b} = 2(q+q')\left(1, \frac{\mathrm{d}|\vec{b}|}{\mathrm{d}E_b}\frac{\vec{b}}{|\vec{b}|}\right) = \\
&= 2\left[(E_q + E_{q'}) - (\vec{q}+\vec{q}\,')\cdot\frac{\vec{b}}{|\vec{b}|\beta_b}\right],
\end{aligned} \tag{5.12}
$$

where $\beta = |\vec{p}|/E$ is the velocity. Equation (5.11) can then be written as:

$$
J_t = \frac{m^2_{qq'}}{E_{q'}} \cdot 2\left[(Eq + E_{q'}) - (\vec{q}+\vec{q}\,')\cdot\frac{\vec{b}}{|\vec{b}|\beta_b}\right]. \tag{5.13}
$$

The Jacobian for the second change of variables, Eq. (5.10), is simpler:

$$
\begin{aligned}
J_{b\bar{b}} &= \frac{\partial m^2_{b\bar{b}}}{\partial E_{\bar{b}}} = 2(b+\bar{b})\frac{\partial \bar{b}}{\partial E_{\bar{b}}} = 2(b+\bar{b})\left(1, \frac{\mathrm{d}|\vec{\bar{b}}|}{\mathrm{d}E_{\bar{b}}}\frac{\vec{\bar{b}}}{|\vec{\bar{b}}|}\right) = \\
&= 2\left(E_b - \vec{b}\cdot\frac{\vec{\bar{b}}}{|\vec{\bar{b}}|\beta_{\bar{b}}}\right).
\end{aligned} \tag{5.14}
$$

After inverting Eqs. (5.13) and (5.14), the complete phase-space volumes for the $\mathrm{t\bar{t}H}$ (S) and $\mathrm{t\bar{t}+b\bar{b}}$ (B) hypotheses are:

$$
\begin{aligned}
\mathrm{d}\Phi_\mathrm{S} = &\left(\frac{1}{2(2\pi)^3}\right)^8 \prod_{i=1}^{2}\left[|\vec{q}_i||\vec{q}_i\,'||\vec{b}_i||J_{t_i}|^{-1}\mathrm{d}E_{q_i}\mathrm{d}m^2_{q_iq_i'}\mathrm{d}m^2_{q_iq_i'b_i}\mathrm{d}\Omega_{q_i}\mathrm{d}\Omega_{q_i'}\mathrm{d}\Omega_{b_i}\right]\times \\
&\times |\vec{b}||\vec{\bar{b}}||J_{b\bar{b}}|^{-1}\mathrm{d}E_b\mathrm{d}m^2_{b\bar{b}}\mathrm{d}\Omega_b\mathrm{d}\Omega_{\bar{b}},
\end{aligned} \tag{5.15}
$$

$$
\begin{aligned}
\mathrm{d}\Phi_\mathrm{B} = &\left(\frac{1}{2(2\pi)^3}\right)^8 \prod_{i=1}^{2}\left[|\vec{q}_i||\vec{q}_i\,'||\vec{b}_i||J_{t_i}|^{-1}\mathrm{d}E_{q_i}\mathrm{d}m^2_{q_iq_i'}\mathrm{d}m^2_{q_iq_i'b_i}\mathrm{d}\Omega_{q_i}\mathrm{d}\Omega_{q_i'}\mathrm{d}\Omega_{b_i}\right]\times \\
&\times |\vec{b}||\vec{\bar{b}}|\mathrm{d}E_b\mathrm{d}E_{\bar{b}}\mathrm{d}\Omega_b\mathrm{d}\Omega_{\bar{b}}.
\end{aligned} \tag{5.16}
$$

5.1.2 Kinematic Reconstruction

The quark energies that have been replaced with invariant masses in Eqs. (5.15) and (5.16) can be expressed in terms of the quark direction, the energy and direction of the other daughter particles, and the mass of the mother particle. For two generic daughter particles with 4-momenta c and d, this involves solving a second order equation in the unknown energy:

$$(c+d)^2 = m_{cd}^2 \quad \Rightarrow \quad E_d = E_d(\Omega_d, \Omega_c, E_c, m_{cd}^2, m_c^2, m_d^2). \tag{5.17}$$

For the signal, there are three different representations of c and d:

$$c = q, \quad d = q', \quad m_{cd} = M_W \tag{5.18}$$

$$c = w, \quad d = b, \quad m_{cd} = M_t \tag{5.19}$$

$$c = b, \quad d = \bar{b}, \quad m_{cd} = M_H \tag{5.20}$$

In the case of the decay of the W boson, W $\rightarrow qq'$, the identifications of Eq. (5.18) can be made and the quarks are assumed massless, $m_q = m_{q'} = 0$. The equation to solve and the solution are then:

$$\begin{aligned}
m_{qq'}^2 &= (p_q + p_{q'})^2 && (5.21)\\
\Rightarrow M_W^2 &= (E_q + E_{q'})^2 - (\vec{p}_q + \vec{p}_{q'})^2 \\
&= E_q^2 + E_{q'}^2 + 2E_q E_{q'} - \vec{p}_q^{\,2} - \vec{p}_{q'}^{\,2} - 2\vec{p}_q \cdot \vec{p}_{q'} \\
&= m_q^2 + m_{q'}^2 + 2E_q E_{q'} - 2|\vec{p}_q||\vec{p}_{q'}| \cos\theta_{qq'} \\
&= 2E_q E_{q'}(1 - \cos\theta_{qq'}) \\
\Rightarrow E_{q'} &= \frac{M_W^2}{2E_q(1-\cos\theta_{qq'})} = \frac{M_W^2}{4E_q \sin^2(\theta_{qq'}/2)}. && (5.22)
\end{aligned}$$

In the case of the decay of the top quark, t $\rightarrow$ Wb, the identifications in (5.19) hold and the bottom quark is assumed massive $M_b > 0$. In order to express the solutions, the following definitions come in handy:

$$a \equiv \frac{M_t^2 - M_W^2 - M_b^2}{2E_w} \tag{5.23}$$

$$b \equiv \cos\theta_{wb} \tag{5.24}$$

$$a' = \frac{a}{M_b} \tag{5.25}$$

$$b' = b|\vec{\beta}_w| \tag{5.26}$$

$$D = a'^2 + b'^2 - a'^2 b'^2 - 1 \tag{5.27}$$

The equation to solve and the solution are then:

$$
\begin{aligned}
m_{wb}^2 &= (p_w + p_b)^2 \qquad (5.28)\\
\Rightarrow M_t^2 &= (E_w + E_b)^2 - (\vec{p}_w + \vec{p}_b)^2\\
&= E_w^2 + E_b^2 + 2E_w E_b - \vec{p}_w^{\,2} - \vec{p}_b^{\,2} - 2\vec{p}_w \cdot \vec{p}_b\\
&= M_W^2 + M_b^2 + 2E_w E_b - 2|\vec{p}_w||\vec{p}_b| \cos\theta_{wb}\\
\Rightarrow \frac{M_t^2 - M_W^2 - M_b^2}{2E_w} &= E_b - \frac{|\vec{p}_w||\vec{p}_b| \cos\theta_{wb}}{E_w}\\
\Rightarrow a &= E_b - \frac{|\vec{p}_w|}{E_w} \cos\theta_{wb} \sqrt{E_b^2 - M_b^2}\\
&= E_b - b' \sqrt{E_b^2 - M_b^2}\\
\Rightarrow (E_b - a)^2 &= b'^2 (E_b^2 - M_b^2)\\
\Rightarrow 0 &= E_b^2(1 - b'^2) - E_b(2a) + (b'^2 M_b^2 + a^2)\\
\Rightarrow E_b &= \frac{2a \pm \sqrt{4a^2 - 4(1 - b'^2)(b'2M_b^2 + a^2)}}{2(1 - b'^2)}\\
&= \frac{a \pm \sqrt{a^2 - (a^2 + b'^2 M_b^2 - b'^4 M_b^2 - b'^2 a^2)}}{1 - b'^2}\\
&= \frac{a \pm \sqrt{-b'^2 M_b^2 + b'^4 M_b^2 + b'^2 a^2}}{1 - b'^2}\\
\Rightarrow E_b^{\pm} &= M_b \frac{a' \pm |b'| \sqrt{a'^2 + b'^2 - 1}}{1 - b'^2}, \qquad (5.29)
\end{aligned}
$$

provided that $E_b^{\pm} > M_b$. The two solutions of (5.29) are not necessarily compatible with Eq. (5.28). The different cases are:

$$
\begin{aligned}
&b' > 0 \text{ and } D < 0 \Longrightarrow E_b = E_b^+ \text{ or } E_b = E_b^-\\
&b' > 0 \text{ and } D > 0 \Longrightarrow E_b = E_b^+\\
&b' < 0 \text{ and } D > 0 \Longrightarrow E_b = E_b^-\\
&b' < 0 \text{ and } D < 0 \Longrightarrow E_b = \varnothing \qquad (5.30)
\end{aligned}
$$

In the case where two solutions are possible, the nearest to the corresponding jet energy is taken, while if no solution exists, the phase space point is declared invalid. This generally only occurs for some but not all of the considered permutations of jet-quark matching in an event.

5.2 Production and Decay Amplitude

The amplitude of the full $\mathrm{t\bar{t}H}$ ($\mathrm{H} \rightarrow \mathrm{b\bar{b}}$) and $\mathrm{t\bar{t}} + \mathrm{b\bar{b}}$ processes, including the subsequent decay of the top quarks, is composed of two different elements. First there is the scattering amplitude of the core $\mathrm{t\bar{t}H}$ or $\mathrm{t\bar{t}} + \mathrm{b\bar{b}}$ processes. Then there are the decay amplitudes of the top quark, W boson and Higgs boson. The forms of the squared signal (S) and background (B) amplitudes are as follows:

$$|\mathcal{M}_{\mathrm{S}}|^2 = \left|\mathcal{M}_{\mathrm{gg}\rightarrow\mathrm{t\bar{t}H}}\right|^2 \cdot \left|\mathcal{M}_{\mathrm{t}\rightarrow\mathrm{qqb}}\right|^2 \cdot \left|\mathcal{M}_{\mathrm{\bar{t}}\rightarrow\mathrm{qqb}}\right|^2 \cdot \left|\mathcal{M}_{\mathrm{H}\rightarrow\mathrm{b\bar{b}}}\right|^2, \tag{5.31}$$

$$|\mathcal{M}_{\mathrm{B}}|^2 = \left|\mathcal{M}_{\mathrm{gg}\rightarrow\mathrm{t\bar{t}b\bar{b}}}\right|^2 \cdot \left|\mathcal{M}_{\mathrm{t}\rightarrow\mathrm{qqb}}\right|^2 \cdot \left|\mathcal{M}_{\mathrm{\bar{t}}\rightarrow\mathrm{qqb}}\right|^2. \tag{5.32}$$

Both amplitudes are discussed in the following sections. Before that, it is useful to define the total 4-momentum of the system:

$$P^\mu = p^\mu_{q_1} + p^\mu_{q'_1} + p^\mu_{b_1} + p^\mu_{q_2} + p^\mu_{q'_2} + p^\mu_{b_2} + p^\mu_{b} + p^\mu_{\bar{b}}. \tag{5.33}$$

5.2.1 *Scattering Amplitude*

The hard scattering amplitude is calculated at LO by the program `OpenLoops` [6]. It is written in `Fortran` and interfaced with the custom `C++` code used for the calculation of the matrix element probability densities. It is called for each integration point and is passed the phase space point consisting of the energy, momentum and mass of the incoming and outgoing particles, and the process identifier, and returns the amplitude squared. The momenta and masses of the outgoing particles are given by:

$$\vec{p}_{t_i} = \vec{p}_{q_i} + \vec{p}_{q'_i} + \vec{p}_{b_i} \qquad \text{and } M_t = 174.3\,\text{GeV} \tag{5.34}$$

$$\vec{p}_H = \vec{p}_b + \vec{p}_{\bar{b}} \qquad \text{and } M_H = 125\,\text{GeV} \tag{5.35}$$

$$\vec{p}_{b,\bar{b}} = \vec{p}_{b,\bar{b}} \qquad \text{and } M_b = 0. \tag{5.36}$$

Since `OpenLoops` requires a balanced system for the LO parton configuration, a Lorentz transformation in the transverse plane is applied to these momenta with the boost vector $\vec{P}_\mathrm{T}/P^0$, such that:

$$(\vec{p}_{t_2})_\mathrm{T} + (\vec{p}_{t_2})_\mathrm{T} + (\vec{p}_H)_\mathrm{T} = (0, 0)$$

$$(\vec{p}_{t_2})_\mathrm{T} + (\vec{p}_{t_2})_\mathrm{T} + (\vec{p}_b)_\mathrm{T} + (\vec{p}_{\bar{b}})_\mathrm{T} = (0, 0)$$

The momenta of the massless incoming gluons, g_i with $i = 1, 2$ are then given by:

$$\vec{p}_{g_i} = (0, 0, \pm(P^0 \pm P^3)/2), \tag{5.37}$$

where P^0 and P^3 are the total energy and net z-momentum of the outgoing particles, respectively. Details of the specific processes used are given below.

$t\bar{t}H$ Signal

The polarisation and spin-averaged scattering amplitude for the $\mathrm{t\bar{t}H}$ signal is given by:

$$|\mathcal{M}_{\mathrm{gg}\to\mathrm{t\bar{t}H}}|^2 = \overline{\sum_{r_1,r_2,s,\bar{s}}} m_{\mathrm{S}}^2(g_1(r_1), g_2(r_2), t(s), \bar{t}(\bar{s}), h)\delta(g_1 + g_2 - t - \bar{t} - h), \tag{5.38}$$

where g_1, g_2 and r_1, r_2 are the 4-momenta and polarisations of the gluons, $t, \bar{t}$ and $s, \bar{s}$ are the 4-momenta and spin states of the top quarks, h is the 4-momenta of the Higgs boson, and the inclusion of a delta function is merely to clarify that the LO amplitudes are defined only for a Born-like configuration.[2] The process used is the $\mathrm{gg} \to \mathrm{t\bar{t}H}$ subprocess of $\mathrm{pp} \to \mathrm{t\bar{t}H}$, which includes all eight LO diagrams as described in Sect. 2.5.1.

$t\bar{t} + b\bar{b}$ Background

The scattering amplitude for the $\mathrm{t\bar{t}} + \mathrm{b\bar{b}}$ background is given by:

$$|\mathcal{M}_{\mathrm{gg}\to\mathrm{t\bar{t}b\bar{b}}}|^2 = \overline{\sum_{r_1,r_2,s,\bar{s}}} m_{\mathrm{B}}^2(g_1, g_2, t, \bar{t}, b, \bar{b})\delta(g_1 + g_2 - t - \bar{t} - b - \bar{b}), \tag{5.39}$$

where the average is the same as in Eq. (5.38), the spin of the b quarks is irrelevant, and the delta function again is merely to clarify that the total momentum of the system is zero. The process considered is the $\mathrm{gg} \to \mathrm{t\bar{t}} + \mathrm{b\bar{b}}$ subprocess of $\mathrm{pp} \to \mathrm{t\bar{t}} + \mathrm{q\bar{q}}$, which includes all relevant diagrams.

QCD Multijet Background

In addition to using the $\mathrm{t\bar{t}H}$ signal and $\mathrm{t\bar{t}} + \mathrm{b\bar{b}}$ background matrix elements, I investigated the use of a second background matrix element to target the QCD multijet background. Given the lack of a 2 to 8 process in `OpenLoops`, I tested several different $2 \to 2$, $2 \to 3$ and $2 \to 4$ processes, as listed in Table 5.1. Despite the promise

[2] A Born-like configuration is the leading term in a Born-series expansion, corresponding to a single interaction at each vertex and nowhere else.

Table 5.1 The different matrix element processes investigated to represent the QCD multijet background. In the second row, reconstruction to top quarks and a Higgs boson was assumed and then those particles were entered into the matrix element calculator as gluons

ME process	Reconstructed				Ignored
$gg \to gg$	$b\bar{b} \to g$	$b\bar{b} \to g$			$q\bar{q}\ q\bar{q}$
$gg \to ggg$	$q\bar{q}b \to t \equiv g$	$q\bar{q}\bar{b} \to \bar{t} \equiv g$	$b\bar{b} \to H \equiv g$		
$gg \to gggg$	$q\bar{q} \to g$	$q\bar{q} \to g$	$b\bar{b} \to g$	$b\bar{b} \to g$	
$gg \to bb\bar{b}\bar{b}$	$b \to b$	$\bar{b} \to \bar{b}$	$b \to b$	$\bar{b} \to \bar{b}$	$q\bar{q}\ q\bar{q}$

of such representative processes, the performance, in terms of the area under a ROC curve,[3] of the $t\bar{t}H$ vs. $t\bar{t} + b\bar{b}$ hypothesis was always better than the $t\bar{t}H$ *vs.* QCD hypothesis, and a second background hypothesis was never implemented. The poor performance arises from the incorrect reconstruction of 2, 3 or 4 massless mother particles from 8 massless daughter particles, which must instead be done considering perturbative corrections. The possibility of specially creating a lowest-order $2 \to 8$ gluon-only process in a matrix element generator remains open for future analyses.

5.2.2 *Top Decay Amplitude*

The decay amplitude of the top quark can be broken down into a propagator amplitude and a vertex decay amplitude:

$$\left|\mathcal{M}_{\mathrm{t}\to\mathrm{qqb}}\right|^2 = \mathcal{M}_{\mathrm{BW}}(t) \cdot |\mathcal{M}_{\Gamma}(\mathrm{t} \to \mathrm{qqb})|^2 . \tag{5.40}$$

$\mathcal{M}_{\mathrm{BW}}$ is proportional to the relativistic Breit–Wigner associated with the top quark or antiquark. The narrow-width approximation ($\Gamma \ll M$) is used, which leads to the appearance of a delta function:

$$\mathcal{M}_{\mathrm{BW}}(t) \propto \frac{1}{(p_t^2 - M_t^2)^2 + \Gamma_t^2 M_t^2} \approx \frac{\pi}{M_t \Gamma_t} \delta(p_t^2 - M_t^2). \tag{5.41}$$

The decay amplitude $|\mathcal{M}_{\Gamma}(\mathrm{t} \to \mathrm{qqb})|^2$ for the unpolarised, on-shell decay of the top quark is derived from Ref. [7], starting with:

$$\mathrm{d}\Gamma_t = \frac{M_t g_W^4}{8\pi^3} \frac{x_q(1 - \mu_b - x_q)}{(x_b - \xi)^2 + \gamma^2} \mathrm{d}x_q \mathrm{d}x_b, \tag{5.42}$$

[3] A receiver operating characteristic (ROC) curve shows the diagnostic ability of a binary classifier as its discrimination threshold is varied.

where the following are dimensionless variables:

$$
\begin{aligned}
x_q &= \frac{2p_q p_t}{m_t^2} \quad \left(= \frac{2E_q}{M_t} \text{ if } |\vec{p}_t| = 0\right),\\
x_b &= \frac{2p_b p_t}{m_t^2} \quad \left(= \frac{2E_b}{M_t} \text{ if } |\vec{p}_t| = 0\right),\\
\mu_b &= \frac{M_b^2}{M_t^2},\\
\xi &= \frac{M_t^2 + M_b^2 - M_W^2}{M_t^2},\\
\gamma &= \frac{M_W \Gamma_W}{M_t^2}.
\end{aligned}
\tag{5.43}
$$

In the rest frame of the top quark, Eq. (5.42) can be expressed as:

$$
\mathrm{d}\Gamma_t = \frac{4M_t^3 g_W^4}{8\pi^3} \frac{x_q(1-\mu_b - x_q)}{(k^2 - M_W^2)^2 + M_W^2 \Gamma_W^2} \mathrm{d}E_b \mathrm{d}E_e, \tag{5.44}
$$

where $k = p_t - p_b$. The general formula for the spin-averaged three body decay of a particle with mass M is given by [8]:

$$
\mathrm{d}\Gamma = \frac{1}{(2\pi)^3} \frac{1}{8M} |\mathcal{M}_\Gamma|^2 \mathrm{d}E_1 \mathrm{d}E_3, \tag{5.45}
$$

where E_1 and E_3 are the energies of two of the three decay products in the rest frame of the mother particle. Comparing (5.45) to Eq. (5.44) provides the following expression for the amplitude squared:

$$
|\mathcal{M}_\Gamma|^2 = \frac{32 M_t^4 g_W^4 x_q (1 - \mu_b - x_q)}{(k^2 - M_W^2)^2 + M_W^2 \Gamma_W^2}. \tag{5.46}
$$

In the limit $\Gamma_W / M_W \to 0$, the following relation holds, cf. (5.41):

$$
\frac{1}{(k - M_W)^2 + M_W^2 \Gamma_W^2} \to \frac{\pi}{M_W \Gamma_W} \delta(k^2 - M_W^2), \tag{5.47}
$$

which leads to:

$$
|\mathcal{M}_\Gamma|^2 = \frac{32\pi M_t^4 g_W^4}{M_W \Gamma_W} x_q (1 - \mu_b - x_q) \delta(k^2 - M_W^2). \tag{5.48}
$$

g_W is equivalent to g_2 of Eq. (2.51), and can be expressed in terms of the Fermi coupling constant:

$$g_W^2 = 4\sqrt{2}M_W^2 G_{\mathrm{F}}. \tag{5.49}$$

The entire amplitude squared for each top quark is therefore:

$$\left|\mathcal{M}_{\mathrm{t}\to\mathrm{qqb}}\right|^2 = \frac{\pi}{M_t\Gamma_t}\delta(p_t^2 - M_t^2)\frac{32\pi M_t^4 g_W^4}{M_W\Gamma_W}x_q(1-\mu_b-x_q)\delta(k^2-M_W^2). \tag{5.50}$$

5.2.3 Higgs Decay Amplitude

The decay amplitude for the Higgs boson can be expressed analogously to that of the top quark:

$$\left|\mathcal{M}_{\mathrm{H}\to\mathrm{b\bar{b}}}\right|^2 = \mathcal{M}_{\mathrm{BW}}(H)\cdot\left|\mathcal{M}_\Gamma(\mathrm{H}\to\mathrm{b\bar{b}})\right|^2. \tag{5.51}$$

Similarly to Eq. (5.41), the narrow-width approximation yields:

$$\mathcal{M}_{\mathrm{BW}}(H) \propto \frac{1}{(p_H^2 - M_H^2)^2 + \Gamma_H^2 M_H^2} \approx \frac{\pi}{M_H\Gamma_H}\delta(p_H^2 - M_H^2), \tag{5.52}$$

while the decay amplitude $\left|\mathcal{M}_\Gamma(\mathrm{H}\to\mathrm{b\bar{b}})\right|^2$ is given by writing Eq. (2.62) specifically for b quarks:

$$|\mathcal{M}_\Gamma|^2 = 3g_b^2 M_H^2\left(1 - \frac{4M_b^2}{M_H^2}\right), \tag{5.53}$$

where $g_b = \sqrt{2}m_b/v$ is the Yukawa coupling to the b quark, defined by Eq. (2.39).

We now have all the components to construct $|\mathcal{M}_{\mathrm{S}}|^2$ and $|\mathcal{M}_{\mathrm{B}}|^2$.

5.3 Transfer Functions

The transfer function is the only place in the construction of the MEM probability density where the detector effects are taken into account. It provides the probability of observing the set of measured observables, $\vec{y}$, given the set of true observables $\vec{x}$:

$$W = W(\vec{y}|\vec{x}). \tag{5.54}$$

In the following, the construction of the transfer function as used in the analysis is described. The construction of a general transfer function relevant to all final-state signatures, can be conceived by generalising the specific case below to include all relevant observables. Where relevant, differences with respect to leptons are noted.

5.3.1 Definition

The set of observables required for the matrix element calculation are the quark momenta, while the set of measured observables are:

- the jet directions: $\Omega = (\cos\theta, \phi)$
- the jet transverse momenta: p_{T}
- the missing transverse momentum: $\not{\vec{p}}_{\mathrm{T}}$.

Given the fact that the detector resolution on reconstructed jet energy is much greater than the angular resolution, the simplifying assumption is made that the directions of the quarks are perfectly measured and given by the direction of the associated jet.[4] A further assumption is made that the jet momenta resulting from the quarks are all independent, as is the missing transverse momentum. The transfer function then becomes a product of individual quark-p_{T} transfer functions with delta functions for the quark directions, and the $\not{\vec{p}}_{\mathrm{T}}$ transfer function:

$$W(\vec{y}|\vec{x}) = \left(\prod_{i=1}^{8} \delta(\Omega_i - \Omega_{qi}) T(p_{\mathrm{T}i}|p_{\mathrm{T}qi})\right) \cdot F(\not{\vec{p}}_{\mathrm{T}}|\vec{0}), \tag{5.55}$$

where p_{T_q} and Ω_q are the true p_{T} and direction of the quark that produces the jet, and there is zero $\not{\vec{p}}_{\mathrm{T}}$ at the quark level.

The quark-p_{T} transfer functions are parameterised in terms of a double gaussian:

$$T(p_{\mathrm{T}}|p_{\mathrm{T}_q}) = p_0 \left[0.7 \exp \frac{-(p_{\mathrm{T}} - p_1)^2}{p_2^2} + 0.3 \exp \frac{(p_{\mathrm{T}} - p_3)^2}{(p_2 + p_4)^2}\right], \tag{5.56}$$

where the normalisation parameter p_0 is non-interesting and the parameters $p_1, \ldots, p_4$ are functions of the quark transverse momentum and pseudorapidity:

$$p_n = p_n(p_{\mathrm{T}_q}, \eta_q), \quad n = 1, 2, 3, 4, \tag{5.57}$$

where the η dependence is a second order effect. Equation (5.56) is essentially a probability density function. For a given quark transverse momentum p_{T_q}, the probability of observing a jet at a given pseudorapidity η with a transverse momentum between p_{T} and $p_{\mathrm{T}} + \mathrm{d}p_{\mathrm{T}}$ is given by $T(p_{\mathrm{T}}|p_{\mathrm{T}_q})\mathrm{d}p_{\mathrm{T}}$.

If an event has fewer than four light-flavour jets or four b jets, or if a jet is ignored, it gives rise to a non-reconstructed jet. The corresponding quark is either out of the

[4]A common assumption employed in signatures containing leptons is that the full momenta of electrons and muons are perfectly measured. This is motivated by the fact that in comparison to jets, their energy resolution is negligible.

detector geometrical acceptance, merged with another quark in a single jet, or the measured energy from its corresponding jet is below the reconstructed-jet-energy threshold. The transfer function becomes an acceptance function dependant on the quark p_{T}, which is given by:

$$A(p_{\mathrm{T}q}, \eta) = \begin{cases} 1 & \text{if } |\eta| > \eta^c \text{ or } \min\{\Delta R_j\} < R^c \\ \int_0^{p_{\mathrm{T}}^c} T(p_{\mathrm{T}}|p_{\mathrm{T}q}) \mathrm{d}p_{\mathrm{T}} & \text{otherwise,} \end{cases} \tag{5.58}$$

where η^c and p_{T}^c are the pseudorapidity and transverse momentum thresholds for selected jets, and R^c is the radius of the jet clustering algorithm. The different cases of Eq. (5.58) are understood as follows:

- A quark out of η-acceptance has a 100% probability to not be reconstructed as a jet, and thus is assigned a value of one.
- If the quark is close enough to another quark q', such that they are merged into a single jet, it also has a 100% probability to not be reconstructed as a jet. Its acceptance function is assigned a value of one, while the other quark's transfer function is modified to include its energy: $T(p'_{\mathrm{T}}|p_{\mathrm{T}q'} + p_{\mathrm{T}q})$.
- In the last case, the measured jet p_{T} is less than the jet selection threshold and the acceptance function is given by the cumulative probability density up to the threshold p_{T}^c.

In this analysis, since the possibility of ignoring jets is used, only the first of the three cases above is employed. In case of a non-reconstructed (lost) jet of a quark in acceptance, the transfer function also returns a value of 1.

The transfer function for the missing transverse energy, $\vec{\not{p}}_{\mathrm{T}} = (\not{p}_x, \not{p}_y)$, is parameterised as a bivariate gaussian:

$$F(\vec{\not{p}}_{\mathrm{T}}|\vec{0}) = \frac{1}{2\pi\sigma_x\sigma_y\sqrt{1-\rho^2}} \exp\left[\frac{1}{2(1-\rho^2)}\left(\frac{\not{p}_x^2}{\sigma_x^2} + \frac{\not{p}_y^2}{\sigma_y^2} - \frac{2\rho\not{p}_x\not{p}_y}{\sigma_x\sigma_y}\right)\right], \tag{5.59}$$

where σ_x and σ_y are the $\vec{\not{p}}_{\mathrm{T}}$ resolutions in the x and y axes and ρ is their correlation. For simplicity, the x and y resolutions are assumed to be equal and $|\vec{\not{p}}_{\mathrm{T}}|$ dependant, $\sigma_x = \sigma_y = \sigma(|\vec{\not{p}}_{\mathrm{T}}|)$, and uncorrelated, $\rho = 0$. Equation (5.59) then reduces to a product of single gaussians:

$$F(\vec{\not{p}}_{\mathrm{T}}|\vec{0}) = \frac{1}{2\pi\sigma^2} e^{-\not{p}_x^2/2\sigma^2} e^{-\not{p}_y^2/2\sigma^2}. \tag{5.60}$$

The transverse recoil, defined as the negative sum of the jet transverse momenta (excluding lost jets) minus the measured missing transverse momentum[5]:

$$\vec{\rho}_{\mathrm{T}} = -\sum_j \vec{p}_{\mathrm{T}j} - \vec{\not{p}}_{\mathrm{T}}, \tag{5.61}$$

[5]In the case of leptons in the final state, this should include the lepton transverse momentum.

is not considered in this analysis. However, the total transverse imbalance of the quark system[6]:

$$\vec{P}_{\mathrm{T}} = -\sum_{q=1}^{8} \vec{p}_{\mathrm{T}q}, \tag{5.62}$$

may be considered in the total transfer function. Its logarithm is passed to a single gaussian with fixed mean μ and resolution σ:

$$R(|\vec{P}_{\mathrm{T}}|) = \frac{1}{\sqrt{2\pi\sigma}} e^{-(\log|\vec{P}_{\mathrm{T}}|/\mathrm{GeV}-\mu)^2/2\sigma}, \tag{5.63}$$

which is then multiplied by $W(\vec{y}|\vec{x})$. However, this feature was not used in this analysis, as described in Sect. 5.4.

5.3.2 Determination of Parameters

The functions defining the four parameters $p_1, \ldots, p_4$ in Eq. (5.56) are derived from from simulated signal events. The p_{T} dependence of the parameters (5.57) is continuous, while the less sensitive η dependence is approximated by considering two bins in $|\eta|$, namely $|\eta| < 1.0$ and $1.0 < |\eta| < 2.5$. Moreover, the parameter functions are derived separately for light-flavour ($q =$ d, u, s, c) and heavy-flavour ($q =$ b) quarks. This leads to four different sets of parameters, each of which is parameterised as follows:

$$\begin{aligned} p_1 &= m_1 + n_1 \cdot p_{\mathrm{T}_q} \\ p_2 &= \sqrt{a_2^2 + b_2^2 \cdot p_{\mathrm{T}_q} + c_2^2 \cdot p_{\mathrm{T}_q}^2} \\ p_3 &= m_3 + n_3 \cdot p_{\mathrm{T}_q} \\ p_4 &= \sqrt{a_4^2 + b_4^2 \cdot p_{\mathrm{T}_q} + c_4^2 \cdot p_{\mathrm{T}_q}^2}, \end{aligned} \tag{5.64}$$

where the p_{T} is expressed in GeV and m, n, a, b, and c are constants. The parameters can be interpreted as the physical quantities of energy response μ and energy resolution σ, which in turn depend on the quark-level p_{T} (in GeV):

$$\begin{aligned} p_1, p_3 &\longrightarrow \mu(p_{\mathrm{T}_q}) = m + n \cdot p_{\mathrm{T}_q} \\ p_2, p_4 &\longrightarrow \sigma(p_{\mathrm{T}_q}) = a \oplus b \cdot \sqrt{p_{\mathrm{T}_q}} \oplus c \cdot p_{\mathrm{T}_q}. \end{aligned} \tag{5.65}$$

[6]In the case of leptons in the final state, this should include the lepton and neutrino transverse momentum.

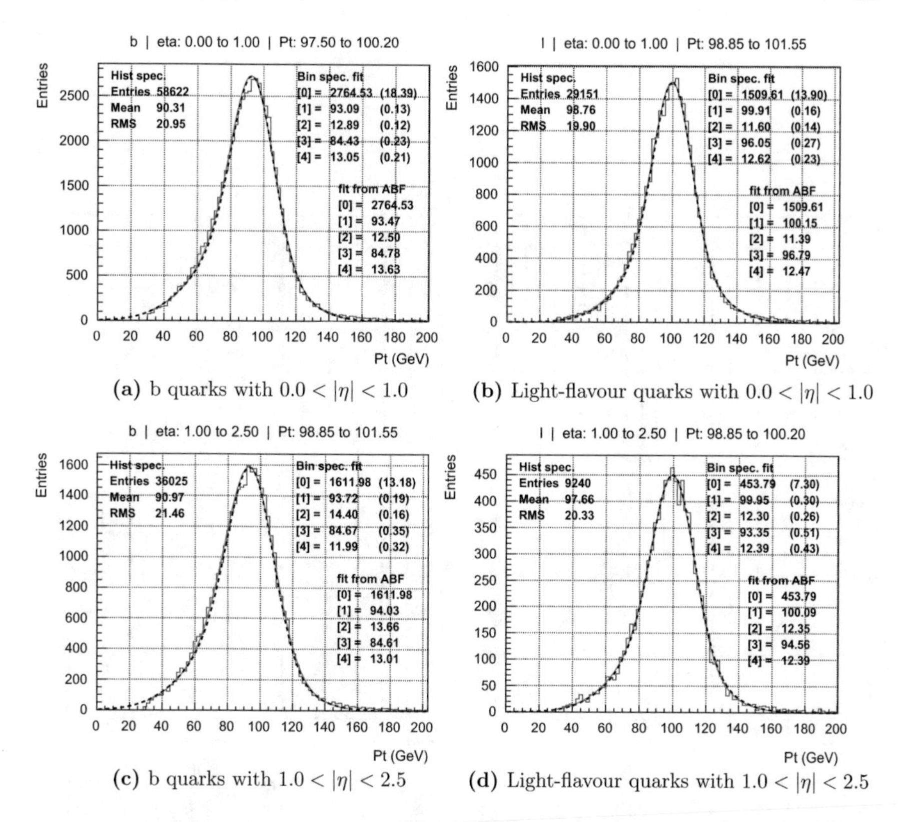

(a) b quarks with $0.0 < |\eta| < 1.0$ (b) Light-flavour quarks with $0.0 < |\eta| < 1.0$

(c) b quarks with $1.0 < |\eta| < 2.5$ (d) Light-flavour quarks with $1.0 < |\eta| < 2.5$

Fig. 5.1 Jet p_{T} (horizontal axis) transfer functions for jets derived from quarks with $p_{\mathrm{T}_q} \sim 100$ GeV for different quark flavours and η bins. The histogram is derived from simulated reconstructed jets arising from quarks with p_{T} in the given range, the smooth line is the double-gaussian (5.56) fitted to this distribution and the black dashed line is the result of using the polynomial fit from Fig. 5.2 for the parameters of the double gaussian. Figures produced by [Joosep Pata, ETH]

In deriving these parameters, individual fits to histograms of reconstructed jet p_{T}, for underlying quark p_{T} in a small window of a few GeV, are made using Eq. (5.56) as the fit function, where the normalisation parameter p_0 is left floating. Examples of these fits for quark p_{T} of around 100 GeV are shown in Fig. 5.1 for light-flavour and b quarks in both bins of $|\eta|$.

A collection of fits similar to those in Fig. 5.1 are made across the quark-p_{T} spectrum from 30 to 300 GeV in variable ranges such that the number of entries (jets) in each range is approximately equal. Then, for each quark-p_{T} bin the four fitted parameters are used in a fit of parameter value *vs.* quark-p_{T}, as shown in Fig. 5.2.

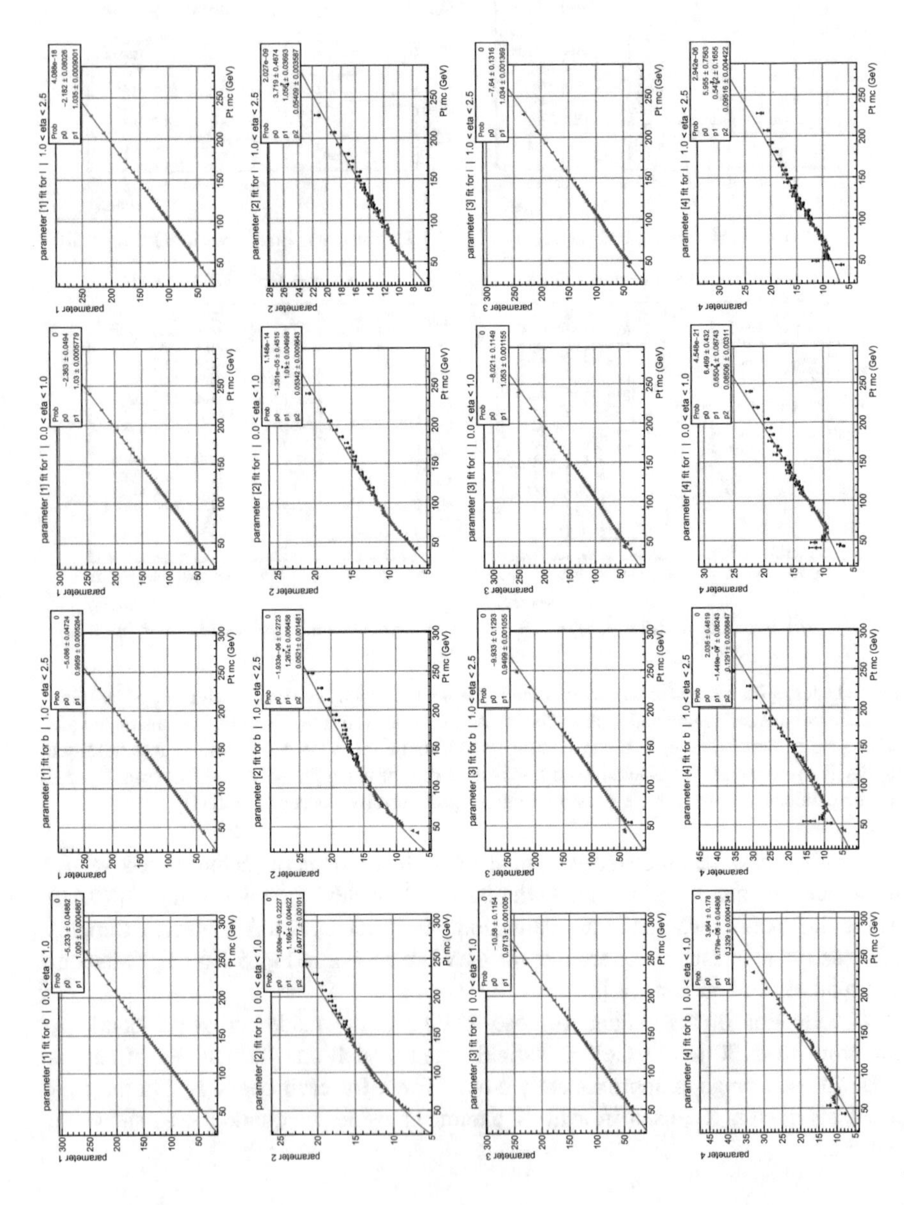

Fig. 5.2 (Continued)

◀ **Fig. 5.2** Transverse momentum dependence of the transfer function fit parameters $p_{1,2,3,4}$. The fit is performed independently in pseudorapidity bins of $0 < |\eta| < 1$ and $1 < |\eta| < 2.5$ and for b- and light-flavoured jets. The slight deviations from the fit in the low end of the p_{T} spectrum in some of the distributions do not appreciably affect the shapes of the transfer functions themselves. Similarly, the single outliers in a few of the plots do not affect the fits. Figures produced by [Joosep Pata, ETH]

The final parameter functions can be determined from the fit results shown in Fig. 5.2. For example, for b quarks with $0 < |\eta| < 1$ they are given by:

$$\begin{aligned} p_1 &= -5.23 + 1.01 \cdot p_{\mathrm{T}_q}, \\ p_2 &= \sqrt{(-1.91 \times 10^{-5})^2 + (1.17)^2 \cdot p_{\mathrm{T}_q} + (0.048)^2 \cdot p_{\mathrm{T}_q}^2}, \\ p_3 &= -10.6 + 0.971 \cdot p_{\mathrm{T}_q}, \\ p_4 &= \sqrt{(3.96)^2 + (9.18 \times 10^{-8})^2 \cdot p_{\mathrm{T}_q} + (0.133)^2 \cdot p_{\mathrm{T}_q}^2}. \end{aligned} \tag{5.66}$$

For the missing transverse momentum transfer function, Eq. (5.60), the resolution parameter $\sigma(|\vec{\not{p}}_{\mathrm{T}}|)$ shows a slight dependency on $p_{\mathrm{T}}^{\mathrm{miss}}$ ranging from around 20 GeV at low $p_{\mathrm{T}}^{\mathrm{miss}}$ (< 100 GeV) to around 40 GeV at high $p_{\mathrm{T}}^{\mathrm{miss}}$ (> 2 TeV). For simplicity however, a constant value of $\sigma = 30$ GeV is used.

5.3.3 Checks and Validation

The derived jet p_{T} transfer functions, Eq. (5.56), are compared for three different values of quark p_{T} and the two $|\eta|$ bins in Fig. 5.3. A slight difference in the transfer functions for $|\eta| < 1$ and $|\eta| > 1$ is observed, thus justifying the coarse η parameterisation. For these functions, and for use in the matrix element, the normalisation parameter p_0 is set to unity, which ensures that jets close to the most probable p_{T} receive the same transfer-function weight irrespective of the quark level p_{T}.

5.4 Event Probabilities

With the key ingredients described thus far, namely the phase space, production and decay amplitudes and the transfer function, the full matrix element probability density can be calculated. For a given assignment of jets to quarks (permutation), and a given process hypothesis i = S, B, the MEM probability density is given by the multidimensional integral (cf. Eq. (2.86)):

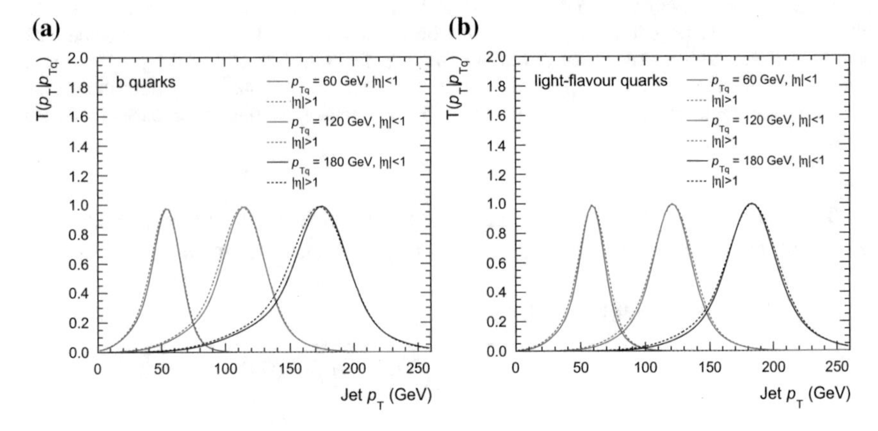

Fig. 5.3 The quark p_{T} transfer function from Eq. (5.56) for three illustrative values of the quark p_{T}, and for b (**a**) and light-flavour (**b**) quarks. The solid lines represent the $0.0 < |\eta| < 1.0$ bin, while the dashed line is for $1.0 < |\eta| < 2.5$

$$w_i(\vec{y}) = \frac{1}{\sigma_i}\int \frac{1}{s}\frac{g(x_a;Q_i)g(x_b;Q_i)}{x_a x_b}\delta^4\Big(x_a P_a + x_b P_b - \sum_{q=1}^{8} p_q\Big)|\mathcal{M}_i(\vec{x})|^2 W(\vec{y}|\vec{x})\mathrm{d}\Phi \mathrm{d}x_a \mathrm{d}x_b, \tag{5.67}$$

where $\sqrt{s}$ is the centre-of-mass pp collision energy, $g(x; Q)$ is the gluon PDF evaluated at the factorisation scale Q, x_a and x_b are the momentum fractions of the initiating partons (gluons), P_a and P_b are the 4-momenta of the colliding protons, which are equal to $P_{a,b} = (\sqrt{s}/2, 0, 0, \pm\sqrt{s}/2)$ in their infinite-momentum frame, and $|\mathcal{M}_i(\vec{x})|^2$, $W(\vec{y}|\vec{x})$ and $\mathrm{d}\Phi$ are defined in Sects. 5.2, 5.3 and 5.1, respectively.

The gluon PDFs are calculated with `LHAPDF` [9] using the `CTEQ6.6` [10] PDF set. The factorisation scale for the signal is fixed at a constant value, $Q_{\mathrm{S}} = m_{\mathrm{t}} + m_{\mathrm{H}}/2$, following Ref. [5], while a dynamical scale, $Q_{\mathrm{B}} = \sqrt{(2m_{\mathrm{t}})^2 + p_{\mathrm{T}_{\mathrm{b}}}^2 + p_{\mathrm{T}_{\bar{\mathrm{b}}}}^2}$, is used for the $\mathrm{t\bar{t}}+\mathrm{b\bar{b}}$ process. The value of $xg(x; Q)$ as a function of Q for different values of x, and as a function of x for different values of Q is shown in Fig. 5.4, using the `CT10nlo` [11] PDF set.

The gluon momentum fractions can be determined from the total 4-momentum of the final-state quark system, $P^{\mu} = \sum_{q=1}^{8} p_q^{\mu}$, via the LO relation, cf. Eq. (5.37):

$$x_{a,b} = \frac{P^0 \pm |P^3|}{\sqrt{s}}. \tag{5.68}$$

The normalisation factor σ_i serves to ensure that w_i is distributed as a probability density, and is defined such that $\int w_i(\vec{y})\mathrm{d}\vec{y} = 1$. In this analysis however, the normalisation of the individual probability densities of the two hypotheses is not imposed, as the relative normalisation is considered in the final discriminant.

The four-dimensional delta function in Eq. (5.67) ensures conservation of 4-momentum. However, in practice, both $\mathrm{t\bar{t}H}$ and $\mathrm{t\bar{t}}+\mathrm{b\bar{b}}$ are not LO processes and they are accompanied by large amounts of initial-state radiation (ISR) and to a lesser extent final-state radiation (FSR). This means the net transverse momentum of the eight final-state particles is seldom zero, and conservation of p_T cannot be expected. On the other hand, conservation of energy and longitudinal momentum is enforced and the delta function is reduced to two dimensions:

$$\delta^4\left(x_a P_a + x_b P_b - \sum_{q=1}^{8} p_q^\mu\right) \to \delta^2(x_a, x_b, P^0, P^3) \cdot \mathcal{R}(\vec{\rho}_\mathrm{T}|\vec{P}_\mathrm{T}), \tag{5.69}$$

where $\mathcal{R}$ is a resolution function which relates the measured p_T imbalance ρ_T to the quark-level p_T imbalance P_T. The two variables are defined in Eqs. (5.61) and (5.62). The resolution function should give higher values for $\vec{\rho}_\mathrm{T}$ close to $\vec{P}_\mathrm{T}$ and smaller values as their difference grows. Therefore, it can be considered a function of $\vec{\rho}_\mathrm{T} - \vec{P}_\mathrm{T} = \vec{\not{p}}_\mathrm{T} + \sum_j \vec{p}_{\mathrm{T}_j} - \sum_q \vec{p}_{\mathrm{T}_q}$. Since the difference between each jet p_T and the corresponding quark p_T is considering in the quark transfer functions, these terms can be neglected. The resolution function is then simply a function of $\vec{\not{p}}_\mathrm{T}$ and is considered in the missing transverse momentum transfer function, Eq. (5.60). The two-dimensional delta function is satisfied by the relation (5.68).

The integration of Eq. (5.67) is performed numerically with the `CUBA` [12] implementation of the `VEGAS` [13] algorithm. After the perfect measurement of quark directions discussed in Sect. 5.3 and the reduction of dimensionality and kinematic reconstruction discussed in Sect. 5.1, the integral reduces to just three dimensions for a fully reconstructed signal event: three one-dimensional integrals over the energy of a light-flavour quark from each top quark decay and a b quark from the Higgs boson decay. In the case of the $\mathrm{t\bar{t}} + \mathrm{b\bar{b}}$ hypothesis an additional dimension enters the integral as the p_T of the second additional b quark. In the case of a lost jet an additional two-dimensional integration over the quark direction is performed. The integration ranges for the variables are as follows:

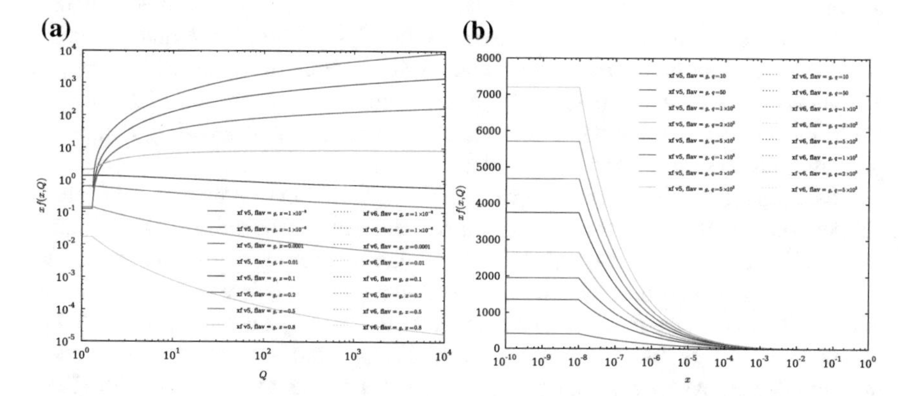

Fig. 5.4 Gluon PDF calculated by `LHAPDF` using the `CT10nlo` PDF set [9]: (**a**) as a function of the factorisation scale Q for different values of the gluon momentum fraction x; (**b**) as a function of x for different values of Q

quark energy, E :	$E \pm 3\sigma_E$
quark polar angle, θ :	$\pm\,\pi/2$
quark azimuthal angle, ϕ :	$[0, 2\pi]$

where σ_E is the jet energy resolution.

5.4.1 Permutations and Hypotheses

The total MEM probability density is then summed over all possible permutations of jet-quark matching, keeping the b and light-flavour quark associations amongst b-tagged and untagged jets. Only inequivalent permutations are considered. For example, interchanging the two light-flavour quarks from the decay of a W boson is considered an equivalent permutation. Similarly, interchanging the two b quarks from the Higgs boson decay or the additional b quarks in the $\mathrm{t\bar{t}} + \mathrm{b\bar{b}}$ process is also equivalent. On the other hand, interchanging a light-flavour quark from one W boson with a quark from the other W boson results in an inequivalent final state, as does interchanging the two b quarks from the top quark decays, and interchanging a b quark from a top quark with a b quark from the Higgs boson or additional radiation.

The number of inequivalent associations of the four light-flavour jets to the four daughters of the W bosons is $3 + 2 + 1 = 6$, while the number of inequivalent b jet to b quark associations is $4 \times 3 = 12$. Therefore, the total number of permutations in a fully reconstructed event with four light-flavour jets and four b tagged jets is $6 \times 12 = 72$. If a light-flavour jet is lost, it is assumed to come from the W^+ boson decay and not permuted, thus the number of permutations is reduced to $3 \times 12 = 36$. If a b jet is lost, it is assumed to come from the top antiquark and is not permuted, resulting in $6 \times 3 = 18$ permutations. The lost b jet is not assumed to come from the Higgs decay as this results in poor discrimination performance, thought to arise from the lack of constraint of the Higgs boson. Additional jets in the event lead to additional permutations where jets are excluded in turn. The b-tagged and untagged jet multiplicities, as well as the number of lost jets considered for the MEM calculation at some point of the analysis, with the corresponding number of permutations, is summarised in Table 5.2.

The choice of which permutations to allow is driven by the matching of jets to quarks in simulation, the computing resources required and the ultimate performance of the MEM discriminant. A first selection of permutations is made considering the matching, where a jet is considered matched to a quark if it is within a distance of $\Delta R = \sqrt{(\Delta\eta)^2 + (\Delta\phi)^2} < 0.3$ to the quark. In the case a jet is within $\Delta R = 0.3$ of two or more quarks, only the closest quark is considered a match. The percentage of simulated signal events which include matches of jets to b quarks from Higgs decays and top decays, as well as to light-flavour quarks from W decays is shown for different jet and b-tag multiplicities in Table 5.3. The corresponding percentages of matches respecting the b flavour status of jets and quarks is shown in Table 5.4.

The choice of requiring b flavour status to match in the allowed permutations is justified given the large number of permutations that would otherwise be required (5 040 in the case of eight jets), and the relatively high matching efficiency shown in Table 5.4 compared to Table 5.3. Excluding some jets from the quark associations in turn can improve the matching efficiency, especially if one or two light-flavour jets are ignored. A study of how many and which jets to ignore was made with regard to computing resource usage and ultimately the MEM discrimination power, and is discussed in Sect. 5.5.1.

The MEM probability density for each permutation is summed to give the total event probability. The idea is that unlikely jet-quark associations will have a very low

Table 5.2 All combinations of the number of jets and b-tagged jets and the relevant assumptions regarding lost jets considered for the MEM calculation as part of testing leading up to the analysis. The corresponding number of permutations for each case is also shown

Jets	b tags	Lost q	Lost b	Permutations
7	3	–	1	$6 \times 3 = 18$
7	3	1	1	$3 \times 3 = 9$
7	4	1	–	$3 \times 12 = 36$
7	4	1	1	$3 \times 3 = 9$
7	4	4	–	$1 \times 12 = 12$
8	3	–	1	$5 \times 6 \times 3 = 90$
8	4	–	–	$6 \times 12 = 72$
8	4	–	1	$4 \times 6 \times 3 = 72$
8	4	1	–	$4 \times 3 \times 12 = 144$
8	4	1	1	$4 \times 4 \times 3 \times 3 = 144$
8	4	4	–	$1 \times 12 = 12$
9	4	–	–	$5 \times 6 \times 12 = 360$
9	4	–	1	$4 \times 5 \times 6 \times 3 = 360$
9	4	4	–	$1 \times 12 = 12$

Table 5.3 The percentage of simulated signal events that include matches of jets to quarks for different jet (*n*j) and b-tag (*n*b) multiplicities. Here *n*Hb means *n* b quarks from the Higgs boson decay are matched to jets. Similarly, *n*Tb means *n* b quarks from the top quark decays are matched, while *n*Wq means *n* light-flavour quarks from the W boson decays are matched to jets. The last four columns show the percentage of events where all eight quarks are matched to jets, all but one b quark from the Higgs decay, all but one b quark from the top quark decays, and all but one light-flavour quark from the W boson decays, respectively

*n*j	*n*b	2Hb (%)	1Hb (%)	2Tb (%)	1Tb (%)	4Wq (%)	3Wq (%)	2Wq (%)	All (%)	1H̸b (%)	1T̸b (%)	1W̸q (%)
7	3	59	38	71	28	9	36	37	–	4	3	12
7	4	75	24	84	16	5	34	40	–	2	1	20
8	3	63	35	73	26	17	38	32	5	5	4	16
8	4	75	23	85	15	16	37	32	8	4	2	23
9	4	75	24	84	16	21	39	29	12	4	3	24

probability and contribute little to the total sum, which is dominated by the "correct" permutation. Theoretically, the sum over permutations can occur inside or outside of the integral in Eq. (5.67) without changing the result. However, in practice, given the finite precision of the numerical integration, different results may arise. The two possibilities were investigated and it was found that summing over the permutations inside the integral resulted in a more accurate integration for a given CPU time. Therefore, Eq. (5.67) must be modified to include the sum over permutations inside the integral. The final MEM probability density for the $\mathrm{t\bar{t}H}$ hypothesis (S) in a fully reconstructed event, after considering the reduction of dimensionality and integrating out the delta functions, is given by:

$$w_{\mathrm{S}}(\vec{y}) = \left(\frac{1}{2(2\pi)^3}\right)^8 \frac{1}{s} \frac{g(x_a; Q_{\mathrm{S}})g(x_b; Q_{\mathrm{S}})}{x_a x_b} \int_0^1 \int_0^1 \int_0^1 \sum_{\text{perm } i} \Big(|\mathcal{M}_{\mathrm{S}}(\vec{x}_i)|^2 W(\vec{y}_i|\vec{x}_i) \times$$

$$\times\, |\vec{q}_{1_i}||\vec{q}\,'_{1_i}||\vec{b}_{1_i}||J_{t_1}|_i^{-1} \cdot |\vec{q}_{2_i}||\vec{q}\,'_{2_i}||\vec{b}_{2_i}||J_{t_2}|_i^{-1} \cdot |\vec{b}_i||\vec{\bar{b}}_i||J_{b\bar{b}}|_i^{-1} \cdot \mathrm{d}E_{q_{1_i}} \mathrm{d}E_{q_{2_i}} \mathrm{d}E_{b_i} \Big), \tag{5.70}$$

where the integration over the energy is converted to a unit scale allowing different energy ranges for each permutation. In the case of the $\mathrm{t\bar{t}} + \mathrm{b\bar{b}}$ hypotheses (B), the factorisation scales Q_{B} and matrix element $\mathcal{M}_{\mathrm{B}}$ are used, an additional integration over $E_{\bar{b}}$ is performed, and the Higgs Jacobian $|J_{b\bar{b}}|$ is removed.

5.4.2 Numerical Integration

The VEGAS algorithm is an iterative and adaptive Monte Carlo integration method which focuses its sampling in regions that make the largest contributions to the integral or to the uncertainty on the integral. It allows multiple iterations of the integral, each of which refines a multi-dimensional search grid from the previous iteration in which to concentrate its sampling. The total estimate of the integral is taken as the error-weighted average of the calculated integral in each iteration. In this analysis up to five iterations of the integral are performed. If the relative precision on

Table 5.4 The percentages of simulated signal events that include matches of jets to quarks for different jet (nj) and b-tag (nb) multiplicities, where the b flavour status of the quarks and jets is also required to match. The columns are defined as in Table 5.3

nj	nb	2Hb (%)	1Hb (%)	2Tb (%)	1Tb (%)	4Wq (%)	3Wq (%)	2Wq (%)	All (%)	1H̸b	1T̸b (%)	1W̸q (%)
7	3	35	57	43	52	5	31	40	–	3	2	–
7	4	66	31	74	24	–	22	44	–	–	–	15
8	3	35	56	40	54	12	35	35	–	5	5	–
8	4	65	32	72	27	7	30	38	6	1	1	17
9	4	62	34	68	30	12	34	35	8	2	1	15

Table 5.5 The number of permutations and function calls per iteration of the integral for each of the five different final states and lost quark hypotheses considered. The average relative precision on the integral and the average CPU time (on an `Intel Xenon E5-2697 v4` processor) for up to five iterations are shown for the $t\bar{t}H$ and $t\bar{t}+b\bar{b}$ hypotheses

Final state	Lost quarks	Number of permutations	No. of calls	Rel. precision		Time (s)	
				$t\bar{t}H$ (%)	$t\bar{t}+b\bar{b}$ (%)	$t\bar{t}H$	$t\bar{t}+b\bar{b}$
7 jets, 3 b tags	1 b	18	4 000	1.3	3.1	46.7	68.4
7 jets, 4 b tags	1 q	36	4 000	2.3	4.3	92.9	136
8 jets, 3 b tags	1 b	90	4 000	1.4	3.4	232	340
8 jets, 4 b tags	1 q	144	4 000	2.4	4.5	373	545
9 jets, 4 b tags	–	360	1 500	1.6	2.8	375	538

the total integral is less than 2% after completing an iteration, no further iterations are attempted.

The maximum number of function calls per iteration is tuned depending on the number of integration variables, so that on average the desired relative precision is achieved. Each function call calculates the MEM probability density, the integrand of Eq. (5.70), in a particular phase space point for each permutation, and thus includes a call to `LHAPDF` for the parton density function, a call to `OpenLoops` for the matrix element calculation, and an evaluation of the transfer function. The total number of integrand evaluations is therefore the number of function calls multiplied by the number of permutations. In order to reduce the number of calls to `OpenLoops`, which is the most time consuming component of the integrand, the matrix element is not calculated when the phase-space point is invalid. The number of function calls, average precision on the integral, and average CPU time employed, for the signal and background hypotheses, for the final jet and b-tag multiplicities and lost quark hypotheses are shown in Table 5.5. The calculation time is proportional to the number of function calls and to the number of permutations.

The integration via `CUBA` allows for the integration of a vector of arbitrary length m, thus making m evaluations of the integrand for each function call. This feature was used in the calculation of systematic uncertainties that affect the reconstructed jets, namely jet energy correction uncertainties (see Sect. 6.6). The typical procedure to propagate a variation in jet p_T to the final discriminant would require the recalculation of the MEM probability densities n times, when n is the number of independent sources of uncertainty. Previously, this number was typically four: the jet energy resolution up and down, and the jet energy scale up and down. In 2016, the jet energy scale uncertainty was factorised into 25 individual sources, thus leading to over 50 systematic variations of the discriminant. These systematic variations form the basis of the vector integrated by `CUBA`.

To avoid over 50 calculations of the matrix element at each function call, the realisation was made that the small variations in jet energy caused by the systematic uncertainties do not appreciably affect the quark-level phase space, as only three jet energies enter the calculation and the integration is performed over three times their jet energy resolution [Joosep Pata, ETH]. In fact, the only non-negligible affect on the MEM probability density from a jet energy variation arises from the transfer function. Therefore each of the 50 plus jet energy variations uses the same matrix element, gluon PDF and phase space calculation, and only the transfer function is calculated n times. This results in a small increase in the total computation time of less than half a percent per systematic variation.

In the case that a jet energy variation changes the selected number of jets or b-tagged jets, a change in MEM hypothesis is evoked which necessitates the recalculation of the full MEM probability density. This occurs in a approximately 10% of simulated events and the recalculated MEM quantities are also made with the vector of jet energy variations, thus requiring only one additional calculation per unique final state. In less than 1% of simulated events, the jet energy variations invoke two additional final states. In these cases, only the first processed additional final state is considered.

5.4.3 *Validation and Performance*

Various checks of the phase-space construction, matrix element evaluation, transfer function values and the convergence of the integral have been performed. The final MEM algorithm described above is stable and robust. The distributions of w_{S} and w_{B} in simulated signal and background events with 8 jets and 4 or more b tags is shown in Fig. 5.5a and b respectively, where one light-flavour quark is assumed to be lost, and only the four most b-like jets are associated with b quarks. The backgrounds shown are the dominant QCD multijet process and the next dominant $\mathrm{t\bar{t}}+$ jets process, split by its various subprocesses, as discussed in Sect. 6.1.2. The discrimination from the individual probability densities is not very good, as can be seen in the ROC curves in the signal *vs.* background efficiency plane of Fig. 5.5c and d, which are constructed by cutting on the variables at different values along their range. The performance can be significantly improved by combining the two probability densities as discussed in the next section.

5.5 Likelihood Discriminant

Given the two probability densities calculated with the MEM for each event, w_{S} for the signal hypothesis and w_{B} for the background hypothesis, a test statistic can be constructed to test these competing hypotheses. According the the Neyman-Pearson lemma [14], the most powerful test statistic to compare two simple hypotheses: the null hypothesis $\mathcal{H}_0$ and a competing hypothesis $\mathcal{H}_1$, is the likelihood ratio:

$$\lambda(X) = \frac{L(X|\mathcal{H}_0)}{L(X|\mathcal{H}_1)}, \tag{5.71}$$

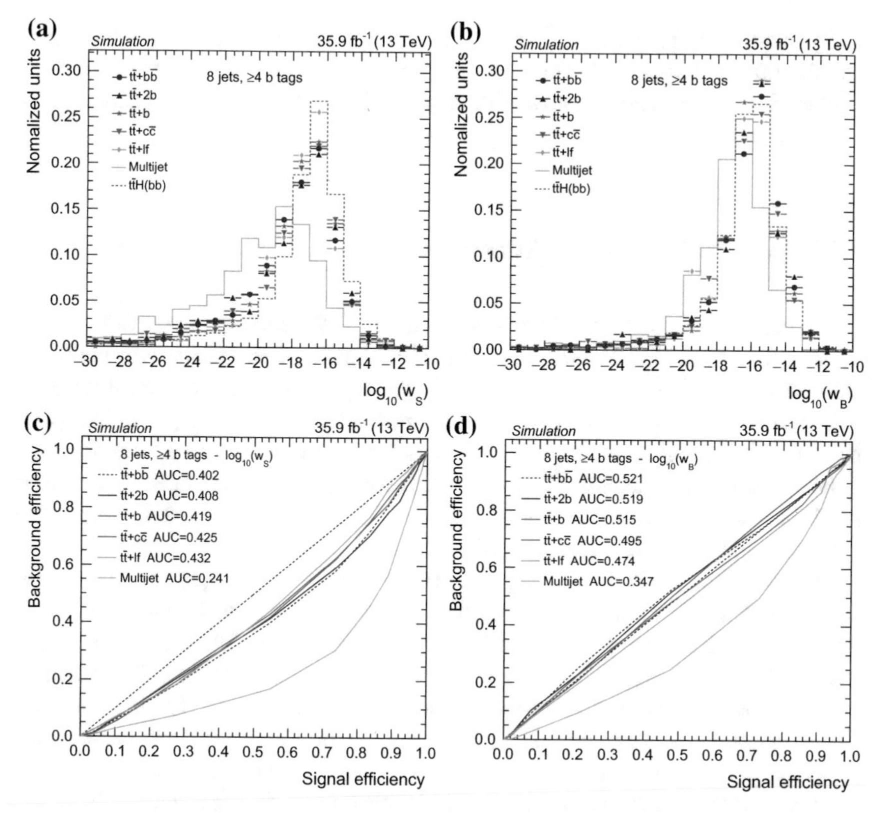

Fig. 5.5 Normalised distributions of w_S (**a**) and w_B (**b**) in simulated events with 8 jets and 4 or more b tags, for the $t\bar{t}H$ ($H \rightarrow b\bar{b}$) signal and $t\bar{t}$ + jets (separated by flavour of the additional quarks) and QCD multijet backgrounds. The corresponding ROC curves are shown in (**c**) for w_S and (**d**) for w_B, along with the area under the curve (AUC) for each background process

where X is the observed data and $L(X|\mathcal{H})$ is the normalised likelihood of observing X under the hypothesis $\mathcal{H}$.

Since no attempt is made to normalise the MEM probability densities, an adjustment to consider the relative normalisations must be made to Eq. (5.71). The optimal discriminator between the signal hypothesis (S) and the background (B) is therefore:

$$\mathcal{P}_{S/B} = \frac{w_S(\vec{y})}{w_S(\vec{y}) + \kappa\, w_B(\vec{y})}, \tag{5.72}$$

where κ is a positive constant that adjusts the relative normalisation. The determination of κ was made to optimise the discrimination power in terms of the expected exclusion limit on the signal strength, described in Sect. 7.1. The optimisation was performed for each unique final state and lost quark hypothesis (category), and started with a value of κ that provides good visual discrimination between the signal and background. Different values of κ above and below its original value were tested until a minimum in the expected limit was found in each category as well as considering combinations of categories. The likelihood ratio $\mathcal{P}_{S/B}$ of Eq. (5.72) is used as the final discriminant in the analysis.

5.5.1 Validation and Performance

The distribution of $\mathcal{P}_{S/B}$ for different values of κ and the corresponding ROC curves are shown in Fig. 5.6, for simulated signal and background events with eight jets and four or more b tags. As can be seen in the figures, the effect of κ on the $\mathcal{P}_{S/B}$ distribution is significant, however the impact on the discrimination power shown by the ROC curve is minimal. Nevertheless, more profound differences were observed in the expected limit, leading to clear choices for the final values of κ. It should be noted that the event selection used in Fig. 5.6 is slightly different to that used in the final analysis,[7] which is described in Sect. 6.3.

As previously mentioned, the possibility to ignore certain jets was tested for each final state of jet and b-tag multiplicity. Specifically, the different assumptions listed in Table 5.2 were tested, and compared in terms of CPU consumption and areas under ROC curves. Ignoring jets requires more integration calls and more permutations, thus increasing the CPU time to beyond acceptable limits in the case of two ignored jets. Ignoring all light-flavour jets on the other hand, reduces the number of permutations, although the lack of constraint led to a poorer performance. The final choice of lost quark hypothesis for each final state is listed in Table 5.6, along with the final optimised values of κ. In this table, the number of jets and b tags refers to the total number observed in the event, however not all of these are considered in the MEM calculation. Specifically, only the best four b-tagged jets are considered as b-quark candidates, while any others are considered light-flavour-quark candidates, and at most five light-flavour-quark-candidates are considered—those with the highest p_T.

[7]The major difference is a selection requirement placed on the quark-gluon discriminator in the analysis and the use of a data-driven estimation method for the QCD multijet background. Other smaller differences are a top quark p_T reweighting and a quark-gluon reweighting, which are not applied here.

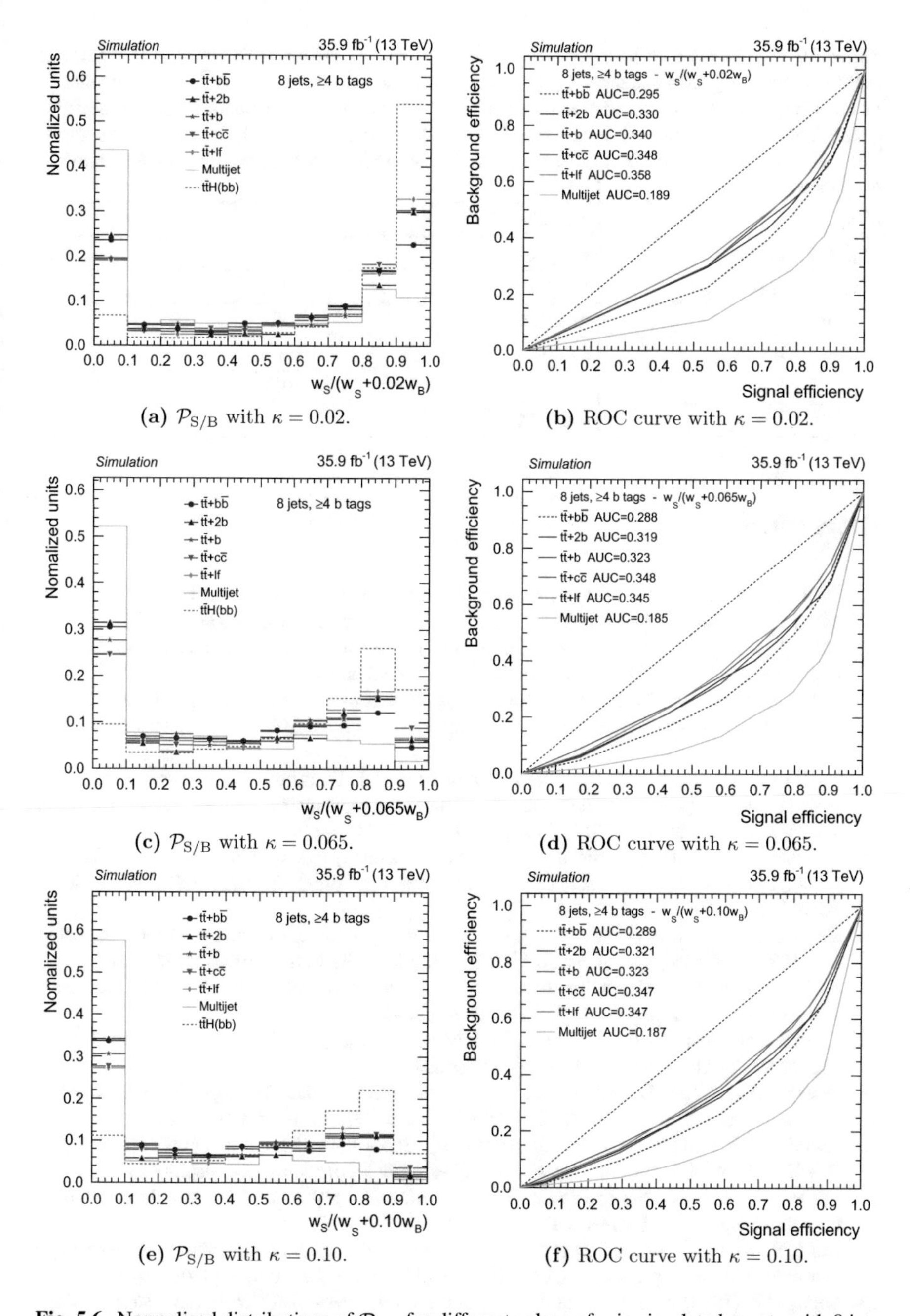

(a) $\mathcal{P}_{\mathrm{S/B}}$ with $\kappa = 0.02$. **(b)** ROC curve with $\kappa = 0.02$.

(c) $\mathcal{P}_{\mathrm{S/B}}$ with $\kappa = 0.065$. **(d)** ROC curve with $\kappa = 0.065$.

(e) $\mathcal{P}_{\mathrm{S/B}}$ with $\kappa = 0.10$. **(f)** ROC curve with $\kappa = 0.10$.

Fig. 5.6 Normalised distributions of $\mathcal{P}_{\mathrm{S/B}}$ for different values of κ in simulated events with 8 jets and 4 or more b tags, for the $\mathrm{t\bar{t}H}$ ($\mathrm{H} \rightarrow \mathrm{b\bar{b}}$) signal and $\mathrm{t\bar{t}}$ + jets (separated by flavour of the additional quarks) and QCD multijet backgrounds. The corresponding ROC curves are shown, along with the area under the curve (AUC) for each background process

Table 5.6 Final choice of lost quark hypothesis per event final state. The `4W2H2T` hypothesis represents the fully reconstructed hypothesis requiring at least 8 jets, `4W2H1T` is the hypothesis where 1 b quark from a top quark is lost, and `3W2H2T` assumes that 1 quark from a W boson is lost. Events with only 3 b tags are assumed to have lost a b quark from the decay of a top quark. The factor κ is used to set the relative normalisation in the final discriminant (5.72), as previously described

Final state	Hypothesis	κ
7 jets, 3 b tags	`4W2H1T`	0.08
8 jets, 3 b tags	`4W2H1T`	0.08
$\geq$9 jets, 3 b tags	`4W2H1T`	0.08
7 jets, $\geq$4 b tags	`3W2H2T`	0.065
8 jets, $\geq$4 b tags	`3W2H2T`	0.065
$\geq$9 jets, $\geq$4 b tags	`4W2H2T`	0.065

References

1. D0 Collaboration (2004) A precision measurement of the mass of the top quark. Nature 429:638–642. https://doi.org/10.1038/nature02589. arXiv:hep-ex/0406031
2. D0 Collaboration (2005) Helicity of the W boson in lepton + jets $t\bar{t}$ events. Phys Lett B 617:1. https://doi.org/10.1016/j.physletb.2005.04.069. arXiv:hep-ex/0404040
3. CDF Collaboration (2009) Search for a Higgs Boson in $WH \rightarrow \ell\nu b\bar{b}$ in $p\bar{p}$ Collisions at $\sqrt{s} = 1.96$ TeV. Phys Rev Lett 103:101802. https://doi.org/10.1103/PhysRevLett.103.101802, arXiv:0906.5613
4. CMS Collaboration (2015) Search for a standard model Higgs boson produced in association with a top-quark pair and decaying to bottom quarks using a matrix element method. Eur Phys J C 75:251. https://doi.org/10.1140/epjc/s10052-015-3454-1, arXiv:1502.02485
5. Dawson S et al (2003) Associated Higgs production with top quarks at the large hadron collider: NLO QCD corrections. Phys Rev D 68:034022. https://doi.org/10.1103/PhysRevD.68.034022. arXiv:hep-ph/0305087
6. Cascioli F, Maierhofer P, Pozzorini S (2012) Scattering amplitudes with open loops. Phys Rev Lett 108:111601. https://doi.org/10.1103/PhysRevLett.108.111601. arXiv:1111.5206
7. Kleiss R, Stirling WJ (1988) Top quark production at hadron colliders: some useful formulae. Z Phys C 40:419. https://doi.org/10.1007/BF01548856
8. Particle Data Group Collaboration (2016) Review of particle physics. Chin Phys C 40:100001. https://doi.org/10.1088/1674-1137/40/10/100001
9. Buckley A et al (2015) LHAPDF6: parton density access in the LHC precision era. Eur Phys J C 75:132. https://doi.org/10.1140/epjc/s10052-015-3318-8. arXiv:1412.7420
10. Nadolsky PM et al (2008) Implications of CTEQ global analysis for collider observables. Phys Rev D 78:013004. https://doi.org/10.1103/PhysRevD.78.013004. arXiv:0802.0007
11. Lai H-L et al (2010) New parton distributions for collider physics. Phys Rev D 82:074024. https://doi.org/10.1103/PhysRevD.82.074024. arXiv:1007.2241
12. Hahn T (2005) CUBA: a library for multidimensional numerical integration. Comput Phys Commun 168:78. https://doi.org/10.1016/j.cpc.2005.01.010. arXiv:hep-ph/0404043
13. Lepage GP (1978) A new algorithm for adaptive multidimensional integration. J Comput Phys 27:192. https://doi.org/10.1016/0021-9991(78)90004-9
14. Neyman J, Pearson ES (1933) On the problem of the most efficient tests of statistical hypotheses. Philos Trans R Soc Lond Ser A 231:289. https://doi.org/10.1098/rsta.1933.0009

Chapter 6
Analysis Strategy

The work presented thus far has been performed with the ultimate aim of searching for $t\bar{t}H$ production in the fully hadronic decay channel. The search has been conducted at CMS and the results have been published in the paper [1]. The details of this analysis are presented below in greater detail than that included in the paper.

6.1 Data and Simulation Samples

In order to determine if detected events arise from the $t\bar{t}H$ process, a thorough understanding of the standard model (SM) background processes that can lead to the same final state is necessary. With such a description, any events observed in excess of SM background expectations can be considered signal events. In this regard, the $t\bar{t}H$ signal and well known background processes are simulated with Monte Carlo (MC) event generators. On the other hand, poorly known background process must be estimated from the data, for example as described in Sect. 6.4.2. The data set used in the analysis and the simulated MC events are described in the following.

6.1.1 Data

The data set analysed was collected from 13 TeV pp collisions by CMS from May to October 2016. Only certified data where all detector subsystems were operating in standard conditions are considered, which corresponds to an integrated luminosity of $35.9\,\text{fb}^{-1}$. The uncertainty on the luminosity measurement is calculated to be 2.5% [2]. The data are recorded in data sets based on trigger types, and for this analysis the `JetHT` data set is used which contains all events selected by any of the jet and H_T based triggers, including the specially developed HLT paths described in Sect. 4.2. The data sets are additionally split by time period which accounts for different beam intensities and other operating conditions. The different data sets

D. Salerno, *The Higgs Boson Produced With Top Quarks in Fully Hadronic Signatures*, Springer Theses, https://doi.org/10.1007/978-3-030-31257-2_6

Table 6.1 The different data sets considered for this analysis as recorded by CMS. The time period of collection, the average instantaneous luminosity and the corresponding integrated luminosity are also shown

Data set	Time period	$\bar{L}$ ($\text{mb}^{-1}\text{s}^{-1}$)	$\mathcal{L}$ (fb^{-1})
JetHT Run B	11 May–21 Jun	4.14	5.75
JetHT Run C	24 Jun–4 Jul	5.87	2.57
JetHT Run D	4 Jul–15 Jul	6.01	4.24
JetHT Run E	15 Jul–25 Jul	6.35	4.02
JetHT Run F	1 Aug–14 Aug	6.80	3.10
JetHT Run G	14 Aug–9 Sep	6.98	7.57
JetHT Run H	25 Sep–26 Oct	7.03	8.65
Total JetHT	11 May–26 Oct	6.05	35.92

used are listed in Table 6.1, along with the time period and corresponding integrated luminosity. Events from these data sets are selected for further analysis if they pass either of the two dedicated signal paths listed in Table 4.1. In addition, to overcome the inefficiency of the H_{T} based triggers in Run H, events passing a single jet trigger are also included in the selection, as described in Sect. 4.2.2.

6.1.2 Simulation Samples

The simulated events used in this analysis are generally produced in three stages. First, there is the simulation of the core physics process and subsequent decays, which is performed by a matrix element event generator as described below. Next is the parton showering and hadronisation of the unstable particles, which is performed by PYTHIA (v.8.2) [3] at leading order (LO). Finally, the simulation of the detector is based on `Geant4 (v.9.4)` [4], which is used to simulate all experimental effects, such as the object reconstruction, selection efficiencies, and detector resolutions.

The $\mathrm{t\bar{t}H}$ signal process is simulated at next-to-leading-order (NLO) with the POWHEG BOX (v.2) [5] event generator. For this simulation, the mass of the Higgs boson is set to $m_{\mathrm{H}} = 125$ GeV and that of the top quark is set to $m_{\mathrm{t}} = 172.5$ GeV. The parton distribution functions (PDFs) of the proton are modelled with `NNPDF3.0` [6]. For targeted optimisation of event selection criteria, the signal is generated in two separate samples, one with the Higgs boson decaying to $\mathrm{b\bar{b}}$, and the other with the Higgs boson decaying to everything except $\mathrm{b\bar{b}}$, namely $\mathrm{W^+W^-}$, gg, $\mathrm{c\bar{c}}$, ZZ, $\gamma\gamma$, $\mathrm{s\bar{s}}$, and $\mu^+\mu^-$.

For the simulation of the background, different event generators are used depending on the process. POWHEG [7–9] is used to simulate the $\mathrm{t\bar{t}}$ + jets and the t- and tW-channels of single top production (single t) at NLO. Associated production of $\mathrm{t\bar{t}}$ with a vector boson, $\mathrm{t\bar{t}}$ + V, is simulated separately as $\mathrm{t\bar{t}}$ + W and $\mathrm{t\bar{t}}$ + Z at NLO with

MadGraph5_aMC@NLO [10]. The production of W and Z bosons with additional jets as well as QCD multijet events is simulated at LO using MADGRAPH [11] with MLM matching.[1] The QCD multijet background is ultimately derived from data as described in Sect. 6.4.2, while simulated events are used in preliminary checks and to ensure the self-consistency of the data-driven estimation method. PYTHIA (v.8.2) is also used to generate the underlying event for the diboson processes, simulated at LO separately for WW, WZ, and ZZ.

The simulated events are characterised by a set of parameters related to cut-off energy scales and energy dependence of the underlying interaction, which is referred to as a *tune*. For all simulated events except $\mathrm{t\bar{t}H}$ and $\mathrm{t\bar{t}}$, the underlying event tune PYTHIA CUETP8M1 [13, 14] is used. For the simulation of $\mathrm{t\bar{t}H}$ and $\mathrm{t\bar{t}}$ events, a custom tune CUETP8M2, developed by CMS with an updated α_s for initial-state radiation, to better model the jet-multiplicity spectrum, is employed.

For an accurate comparison with data, the simulated samples need to be normalised to the integrated luminosity of the data according to their predicted cross sections. For the $\mathrm{t\bar{t}}$ simulated events, the cross section is calculated at full next-to-next-to-leading-order (NNLO) accuracy with soft-gluon resummation at next-to-next-to-leading-logarithmic (NNLL) accuracy [15], where the top quark mass is assumed to be $m_\mathrm{t} = 172.5$ GeV, and NNPDF3.0 is used for the PDFs of the proton. For the other backgrounds, the cross sections are calculated at NNLO for W + jets and Z + jets production, approximate NNLO for the single top quark tW and s channels [16], and NLO for the single top quark t channel [17], diboson [18] and $\mathrm{t\bar{t}} + \mathrm{V}$ [19] production. The production cross section of $\mathrm{t\bar{t}H}$ and the Higgs boson branching ratios are also calculated at NLO accuracy [20].

A list of the simulated samples used in the analysis, along with the matrix element generator, production cross section, number of generated events and equivalent integrated luminosity, is provided in Table 6.2.

The additional jets in the $\mathrm{t\bar{t}}$ + jets process originate from different underlying processes and therefore have different kinematic properties and systematic uncertainties. To exploit the different kinematic properties and correctly account for the systematic uncertainties, the events in the $\mathrm{t\bar{t}}$ sample are separated according to the flavour of the additional jets that do not originate from the top quark decays. The identification is made by matching generator-level jets with their originating partons, and results in the following classification:

- $\mathrm{t\bar{t}} + \mathrm{b\bar{b}}$: the event contains two additional b jets, each of which originates from one or more overlapping b quarks.
- $\mathrm{t\bar{t}} + \mathrm{b}$: the event has only one additional b jet which originates from a single b quark.
- $\mathrm{t\bar{t}} + 2\mathrm{b}$: the event contains one additional b jet which originates from two or more overlapping b quarks.

[1]MLM matching [12], named after the original developer Michelangelo L. Mangano, involves matching the final jets after parton-shower evolution and jet clustering to the original partons from the matrix element calculation. This is necessary to avoid double counting jets.

Table 6.2 Generated MC samples used in this analysis with the corresponding cross sections (including branching ratio to the final state where required), number of generated events and equivalent integrated luminosity. The numbers in parentheses indicate the H_T range (in GeV) of the interacting partons for the samples in which this is restricted

Process (pp →)	Generator	$\sigma \cdot$ BR (pb)	Events (M)	$\mathcal{L}$ (fb^{-1})
$t\bar{t}H$, $H \to b\bar{b}$	POWHEG	$0.5071 \cdot 0.5824$	3.7	12 655
$t\bar{t}H$, $H \not\to b\bar{b}$	POWHEG	$0.5071 \cdot 0.4176$	3.8	18 294
$t\bar{t}$ + jets	POWHEG	831.8	77	93
$t\bar{t}$ + W, $W \to q\bar{q}'$	MG5_aMC@NLO	0.4062	0.43	1 056
$t\bar{t}$ + Z, $Z \to q\bar{q}$	MG5_aMC@NLO	0.5297	0.35	655
W + jets, $W \to q\bar{q}'$ (> 180)	MG5_aMC@NLO	2 788	22	8.0
Z + jets, $Z \to q\bar{q}$ (> 600)	MG5_aMC@NLO	5.670	1.0	176
tW	POWHEG	35.85	0.99	28
$\bar{t}$W	POWHEG	35.85	1.0	28
t (t-channel)	POWHEG	136.0	6.0	44
$\bar{t}$ (t-channel)	POWHEG	80.95	3.9	49
t (s-channel)	MG5_aMC@NLO	10.32	1.9	181
WW	PYTHIA8	118.7	8.0	67
WZ	PYTHIA8	47.13	4.0	85
ZZ	PYTHIA8	16.52	2.0	120
QCD multijet (300, 500)	MG5_aMC@NLO	351 300	55	0.16
QCD multijet (500, 700)	MG5_aMC@NLO	31 630	62	2.0
QCD multijet (700, 1000)	MG5_aMC@NLO	6 802	45	6.6
QCD multijet (1000, 1500)	MG5_aMC@NLO	1 206	15	13
QCD multijet (1500, 2000)	MG5_aMC@NLO	108.4	4.0	37
QCD multijet (>2000)	MG5_aMC@NLO	22.72	2.0	87

- $t\bar{t} + c\bar{c}$: the event has at least one additional c jet, each of which originates from one or more overlapping c quarks.
- $t\bar{t}$ + lf: the event is not classified as any of the above.

In addition to the primary processes of interest, effects from pileup are modelled by adding simulated minimum-bias pp events, generated with PYTHIA, to all simulated samples. The number of minimum-bias events added to a particular event is randomly selected according to a predetermined distribution of the pileup multiplicity. This distribution is set to match the expected distribution in data, however since the simulated samples were generated before the data taking was completed, the pileup multiplicity in simulation is different to that observed in the data. A pileup reweighting procedure is therefore applied to MC samples as described in Sect. 6.2.1.

6.2 Event Reweighting

Although every effort is made to generate MC samples that accurately reflect the true physics processes, small discrepancies can often arise. In general, the modelling of important simulated processes is compared to data in control regions where the process of interest contributes the majority, if not all, of the events. Differences observed in these control regions are used to derive corrections based on a particular variable or several variables, which are then applied to the simulated samples for use in all regions. These corrections typically result in a weight for each MC event that is directly related to the over or underestimation of events with the particular properties of the correction. The procedure is referred to as reweighting and is described in the following for all relevant corrections applied in this analysis.

6.2.1 Pileup Reweighting

During the 2016 data taking period, the LHC provided increasingly large instantaneous luminosities to the experiments, as can be seen in Table 6.1. The result of this was an increase of the average rate of overlapping events over time. These pileup events that occur alongside the physics events of interest can impact the object identification and performance, e.g. the lepton isolation or jet reconstruction. Therefore, it is important that the simulated MC events have the same distribution of pileup events as that observed in data.

As already mentioned in Sect. 6.1, the average amount of pileup in 2016 data was unknown at the time of MC event generation. Therefore, the pileup distribution in simulated events must be reweighted to match the observed distribution in data. For example, if 20% of all simulated events have 50 additional pileup events, but only 10% of data events have this number, then simulated events with 50 pileup events must only count for half an event. The actual distribution of the number of pileup events included in simulated events is depicted in Fig. 6.1a.

For the data, the number of pileup interactions for each collision depends on the instantaneous luminosity for each bunch crossing and the total inelastic cross section, $\sigma_{\text{inelastic}}$. The instantaneous luminosity of each bunch crossing is estimated based on the average and RMS values of the instantaneous luminosity per bunch crossing for a given luminosity section. The total inelastic pp cross section is measured using dedicated forward detectors and $\sigma_{\text{inelastic}} = 69.2$ mb is found to accurately describe the 13 TeV data, with an uncertainty of 4.6% [21]. The resulting target pileup distribution expected in data, together with the varied distributions obtained by varying $\sigma_{\text{inelastic}}$ up and down by 4.6%, are shown in Fig. 6.1b. The two distributions in Fig. 6.1 are used to compute a pileup weight for each simulated event based on the number of pileup events, as well as the corresponding systematic variations.

The effect of the pileup reweighting can be seen in the distribution of the number of reconstructed vertices, which is correlated with the number of pileup events.

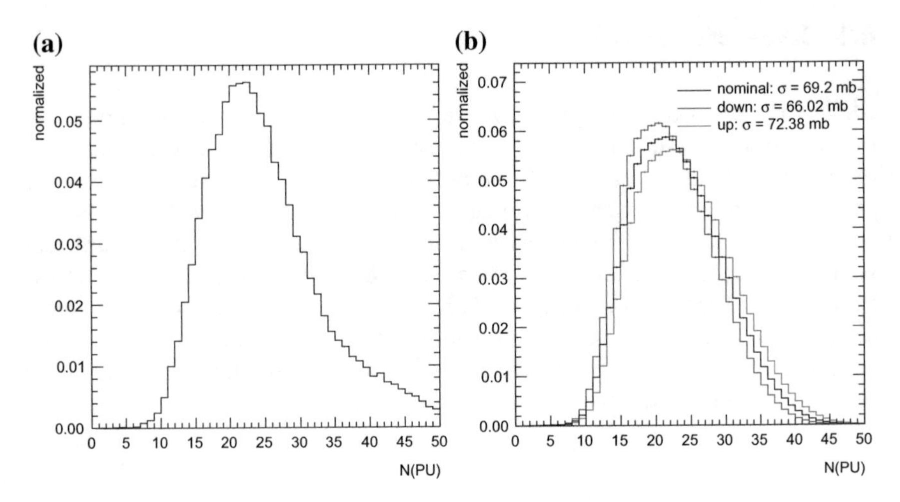

Fig. 6.1 (**a**) Distribution of the number of pileup (PU) events in simulated events. (**b**) Target distribution for the analysed data calculated using information about the instantaneous luminosity and a total inelastic pp scattering cross section of $\sigma_{\text{inelastic}} = 69.2\,\text{mb}$ (central histogram). The distributions obtained when varying $\sigma_{\text{inelastic}}$ by its uncertainty of 4.6% are also shown

Figure 6.2 shows the number of reconstructed primary vertices in data and in simulated events in the preselection region (see Sect. 6.3.1), before and after the pileup reweighting. The imperfect agreement after the reweighting is attributed to the non-ideal modelling of vertex-related quantities. Tests using different pileup reweighting scenarios show that other distributions show no dependency on the pileup reweighting. In addition, as will be shown later in Sects. 6.3.1, 6.4.2, and 6.5, variables used to separate the signal from the background are well modelled by the simulation and are not affected by this mismodelling of the vertex distribution.

6.2.2 *Top p_T Reweighting*

One of the most important processes to simulate is that of $t\bar{t}$ production. Although the simulation of $t\bar{t}$ is rather accurate and improving over time, some differences still exist in the spectra of variables from different MC generators, both amongst themselves and compared to data. A particularly noteworthy discrepancy is observed in the p_T distribution of the top quark, which is shown by differential $t\bar{t}$ cross section measurements to be considerably harder in simulation than data [22]. A reweighting is therefore applied to the POWHEG generated $t\bar{t}$ events used in this analysis, as a function of the generator level p_T of the top quarks.

The reweighting function has been derived from the differential cross section of semi-leptonic $t\bar{t}$ production with zero additional jets, shown in Fig. 6.3a. The requirement of zero additional jets ensures that the $t\bar{t}H$ signal is not included in

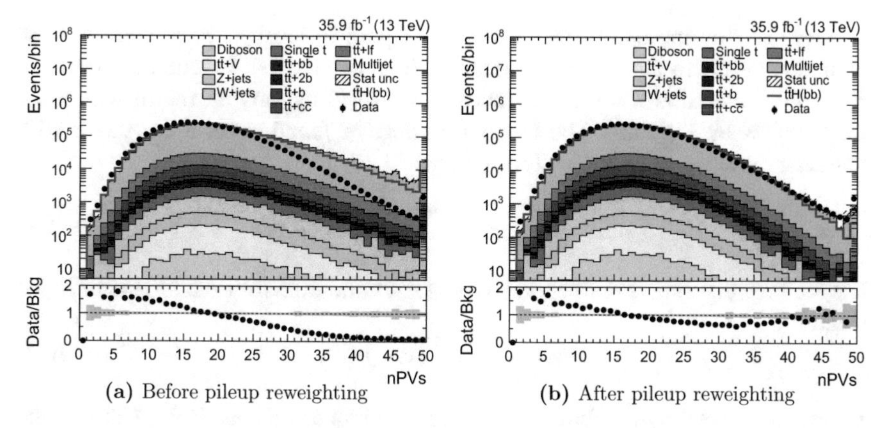

(a) Before pileup reweighting (b) After pileup reweighting

Fig. 6.2 Comparison of number of reconstructed vertices (nPVs) in data (black markers) and in simulation (stacked histograms) before and after applying the pileup reweighting, after the preselection. The simulated backgrounds are scaled to the luminosity of the data and then the QCD multijet background is further scaled to match the yield in data. The signal contribution is scaled to the total background yield (equivalent to the data yield) for better readability. The uncertainty bands include statistical uncertainties only

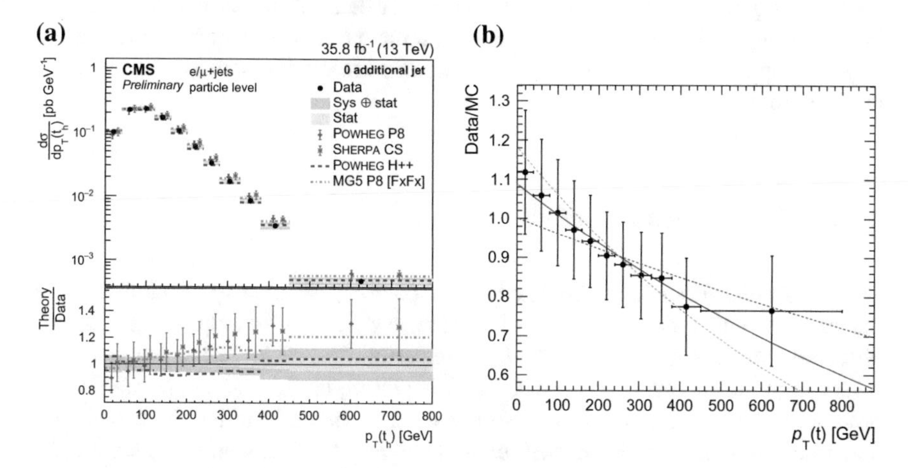

Fig. 6.3 (**a**) The particle-level top quark p_T distribution in the zero additional jet region of the semi-leptonic top-quark decay channel [22]. (**b**) The ratio of data to MC simulation for POWHEG+PYTHIA8 generated events. The best-fitted exponential function is shown as the blue line, while the up and down systematic variations are shown as the red and green dashed lines

the cross section measurements. The ratio of particle-level p_T in POWHEG+PYTHIA8 simulated events to that in data is fitted with an exponential function as shown in Fig. 6.3b. Uncertainties on this function are derived by varying the fit within the uncertainties on the ratio. The nominal correction function for a top quark with generator-level $p_T = x$ GeV is derived to be:

$$f(x) = e^{0.08436 - 0.00074\,x}. \tag{6.1}$$

The total reweighting is defined in terms of the correction for each top quark in the event:

$$w_{\text{top}\,p_T} = \sqrt{f(x) \cdot f(\bar{x})}, \tag{6.2}$$

where x and $\bar{x}$ are the p_T of the top quark and top antiquark, respectively. Although the correction is derived from particle-level measurements, it is applied to the generator-level p_T. This is justified as the parton-level and particle-level p_T of the inclusive selection regions in Ref. [22] are very similar. The effect of this reweighting can be seen in Fig. 6.4, which shows the distribution of several event variables in data and simulated events in a single-muon $t\bar{t}$ validation region (see Sect. 6.4.1), before and after the top p_T correction. The agreement between data and the predicted background is clearly improved, however discrepancies remain due to the imperfect modelling of jet multiplicity. These differences are accounted for with systematic uncertainties assigned to the $t\bar{t}$ simulation.

6.2.3 Trigger Scale Factors

The trigger performance in simulated events does not necessarily match the performance observed in data. Initially, a decrease in efficiency in data at high H_T was observed, which is attributed to the last run period of the LHC (Run H) which had a very high instantaneous luminosity. In this period, the L1 H_T triggers suffered a problem in which saturated (high p_T) jets were excluded from the H_T calculation, thus resulting in a much lower measurement of H_T. As discussed in Sect. 4.2.2, the OR of a single-jet trigger with a p_T threshold of 450 GeV is included, which ensures that events with high p_T jets are selected.

The overall trigger efficiency is calculated in a single pass, as a function of the number of offline b-tagged jets (CSVM), the p_T of the 6th jet and the event H_T, by using a single muon data set collected with a single muon trigger and comparing the number of events passing the signal triggers to the total number of selected events. For an accurate comparison, the trigger efficiency in simulation with respect to the offline selection is measured in the same way, and by applying the same single-muon trigger. The trigger efficiencies as a function of the p_T of the 6th jet, the H_T and the number of b-tagged jets are shown in Fig. 6.5a–c. The benefit of the high-H_T inefficiency mitigation strategy can be seen by comparing Fig. 6.5b and d, which

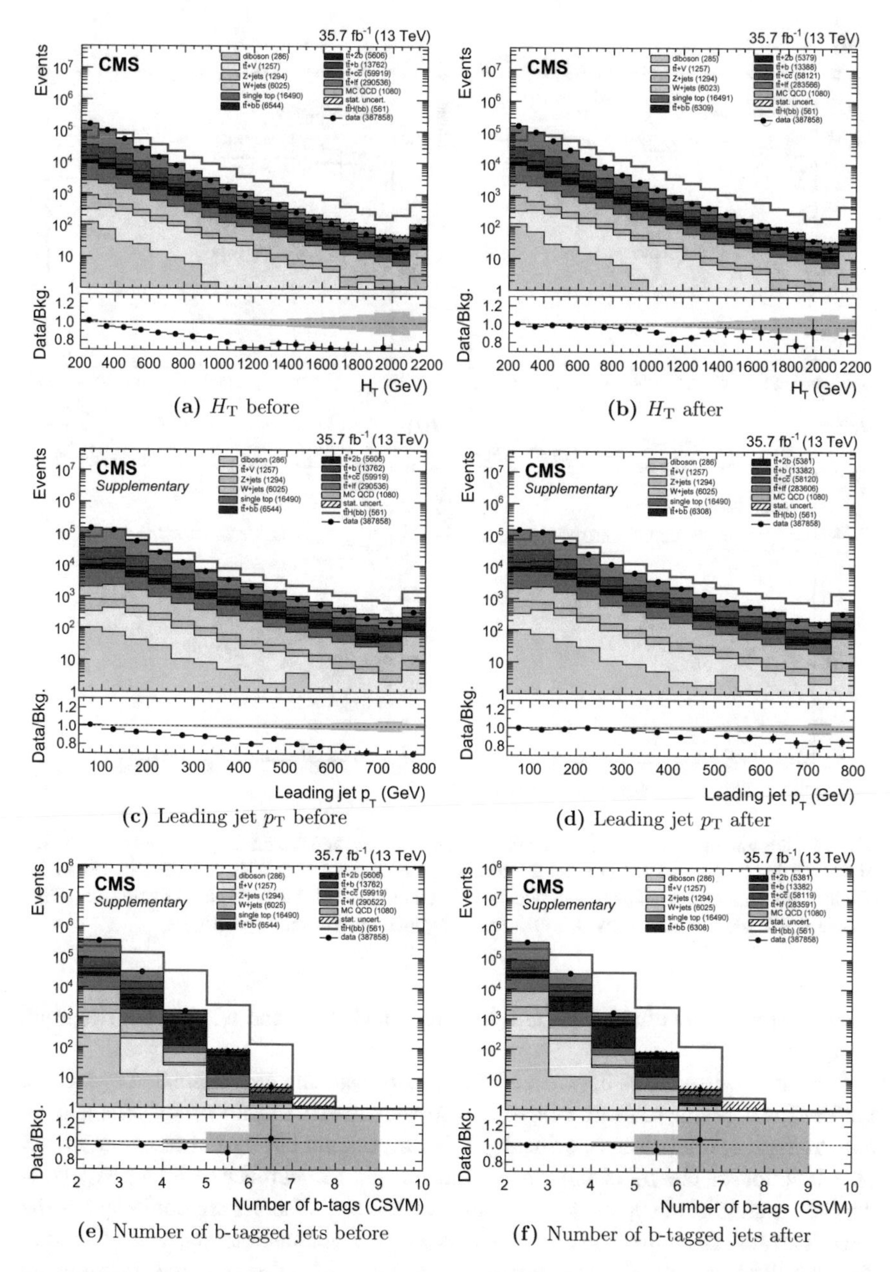

(a) H_T before (b) H_T after

(c) Leading jet p_T before (d) Leading jet p_T after

(e) Number of b-tagged jets before (f) Number of b-tagged jets after

Fig. 6.4 Distributions in data and simulated events (dominated by $\mathrm{t\bar{t}}$) in a single-muon validation region (see Sect. 6.4.1), before and after applying the top p_T reweighting. The simulated backgrounds are scaled to the luminosity of the data. The uncertainty bands include statistical uncertainties only

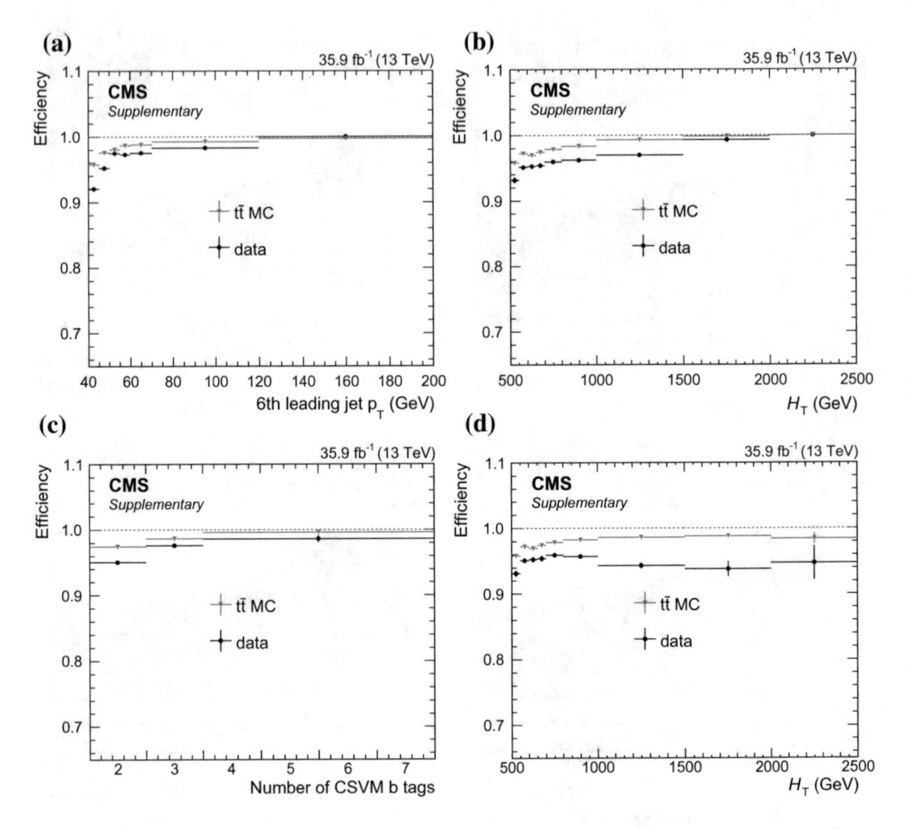

Fig. 6.5 Trigger efficiencies of the OR of both signal triggers and the single-jet trigger as a function of the 6th jet p_T (**a**), the HT (**b**) and the number of b-tagged jets (**c**). Events are selected with a single muon trigger as well as the preselection described in Sect. 6.3.1. Jets are selected according to Sect. 4.3.5. (**d**) The efficiency as a function of H_T without the single-jet trigger

show the respective efficiencies as a function of H_T with and without the single-jet trigger.

The slight differences between efficiencies in data and MC simulated events are rectified by applying scale factors, which are calculated as the ratio of the efficiency in data to that in simulation. Specifically, a 3-dimensional bin-by-bin rescaling factor, as a function of the p_T of the 6th jet, the H_T and the number of b-tagged jets, is applied to simulated events. Uncertainties on the scale factors are computed as the uncertainty on the ratios and are treated as a systematic uncertainty, as described in Sect. 6.6. The overall trigger efficiency for signal events that pass the offline event selection is 99.0%, while the derived scale factors are mostly around 0.99, but range from 0.83 to 1.04. The uncertainties on the scale factors are around 1.5% on average with some as high as 15%. The trigger scale factors and uncertainties are reported for a few values of H_T, p_T and b-tag multiplicity in Table 6.3.

Table 6.3 Values of the derived trigger efficiency scale factor and associated uncertainty for selected values of the 6th jet p_T, the H_T and the number of b-tagged jets (nB)

p_{T6}	H_T	nB	Scale factor
40	600	3	0.992 ± 0.017
50	600	3	0.995 ± 0.016
60	600	3	0.974 ± 0.025
50	500	3	1.015 ± 0.009
50	600	3	0.995 ± 0.016
50	700	3	1.010 ± 0.013
50	600	2	0.989 ± 0.009
50	600	3	0.995 ± 0.016
50	600	≥ 4	1.000 ± 0.000

6.2.4 B-Tagging Scale Factors

In general, the b-jet identification efficiency and the misidentification probability of the b-tagging algorithm described in Sect. 4.3.6 differs in data and simulation. Therefore, the distribution of the b-tagging discriminant in simulation is corrected by scale factors, which depend on the flavour, p_T and $|\eta|$ of the jets [23], to better describe the distribution observed in data. This correction is derived separately for light-flavour and b jets from a "tag-and-probe" approach using control samples enriched in events with a Z boson and exactly two jets, and $t\bar{t}$ events with no additional jets, respectively.

In the absence of a data-driven calibration sample for charm jets, the scale factors for c jets are set to 1.00 and an uncertainty on this scale factor is derived from the calibration for b jets. In the final event selection, described in Sect. 6.3.2, it is estimated that around 62% (44%) of background events have a charm jet among the selected jets (b-tagged jets) of the signal region, while around 67% (22%) of signal events in the same region have a charm jet. A total scale factor is applied to the event, which is calculated as the product of the scale factors of each jet i:

$$\mathrm{SF}_{\mathrm{total}} = \prod_{i=1}^{N_{\mathrm{jets}}} \mathrm{SF}_i = \mathrm{SF}_1 \cdot \mathrm{SF}_2 \cdot \ldots \cdot \mathrm{SF}_{N_{\mathrm{jets}}} \quad (6.3)$$

The systematic uncertainties on the b-tagging scale factors are considered in the final result and described in Sect. 6.6.

The distribution of the b-tagging discriminator variable is shown in Fig. 6.6 for data and simulation before and after applying the b-tagging scale factors.

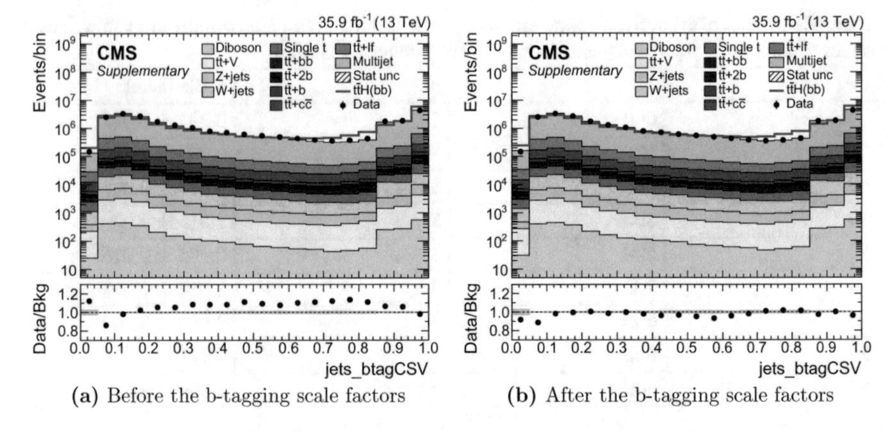

(a) Before the b-tagging scale factors (b) After the b-tagging scale factors

Fig. 6.6 Comparison of the b-tagging (CSV) distribution of all jets in data (black markers) and in simulation (stacked histograms), before and after applying the b-tagging scale factors, after the preselection. The simulated backgrounds are scaled to the luminosity of the data and then the QCD multijet background is further scaled to match the yield in data. The uncertainty bands include statistical uncertainties only

6.2.5 Quark-Gluon Likelihood Reweighting

The distribution of the quark-gluon likelihood (QGL) discriminant for jets, described in Sect. 4.3.7, is different in data and simulated events, and therefore the number of events passing the quark-gluon likelihood ratio (QGLR) selection also differs in data and simulation. To correct for this difference, an event-based reweighting is applied based on the flavour (quark or gluon) and QGL value of all jets in the event [24]. The normalisation impact of the QGL reweighting itself is corrected to ensure the yield after all cuts excluding the cut on QGLR is unchanged. The uncertainty of the reweighting is considered as the full correction, i.e. the up and down variations are taken as no QGL reweighting and twice the reweighting minus one, respectively, as described in Sect. 6.6.

The distribution of the QGLR calculated excluding the first 3 b-tagged jets is shown in Fig. 6.7 for data and simulation for all events passing the preselection. The distributions are shown before and after the QGL reweighting, which clearly improves the agreement between simulation and data.

6.3 Event Selection

For both data and simulated events, an event cleaning procedure is applied to remove events that are either non-physical or uninteresting. Specifically, each event must contain at least one primary vertex (PV) that passes the following selection criteria:

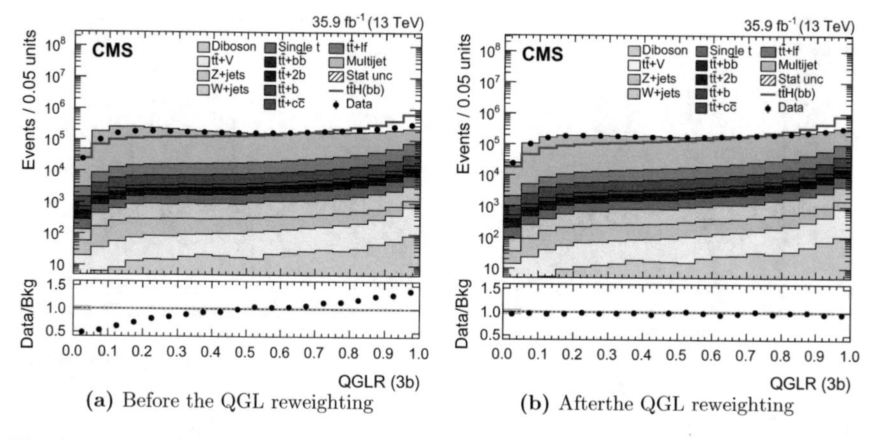

(a) Before the QGL reweighting (b) Afterthe QGL reweighting

Fig. 6.7 Comparison of the quark-gluon likelihood ratio calculated excluding the first 3 b tagged jets in data (black markers) and in simulation (stacked histograms), before and after applying the QGL reweighting, after the preselection. The simulated backgrounds are scaled to the luminosity of the data and then the simulated QCD multijet background is further scaled to match the yield in data. The signal contribution is scaled to the total background yield (equivalent to the data yield) for better readability. The uncertainty bands include statistical uncertainties only

- the number of degrees of freedom used to find the PV must be 5 or more;
- the absolute value of the z-coordinate of the PV must be less than 24 cm;
- the absolute value of the r-coordinate of the PV must be less than 2 cm;
- the PV must be matched to a simulated vertex for simulated events.

Since events in data can only be collected after passing the dedicated triggers, the MC simulated events are required to pass the same triggers using a version of the online reconstruction software, which forms the first stage of the event selection. To overcome trigger inefficiencies near the trigger thresholds, a preselection is made, which forms the second stage of event selection. Events remaining after the preselection are then analysed and an optimal selection is made from which the signal can be extracted. The trigger selection has been discussed in Sect. 4.2, while the preselection and final selection are discussed in the remainder of this section.

6.3.1 Preselection

In order to ensure that analysed events fall in or close to the plateau of the triggers, a preselection based on the same variables used in the triggers is made. The choice of preselection is a delicate balance between maximising the trigger efficiency and minimising the signal loss. In making this choose, the distributions of offline-reconstructed quantities in simulated fully hadronic $t\bar{t}H\,(H \to b\bar{b})$ events were considered. Several such variables, with a sole requirement on selected jets of $p_T > 15$ GeV and $|\eta| < 4.7$, are shown in Fig. 6.8. From these distributions, the following selection criteria that retain most of the signal can immediately be applied:

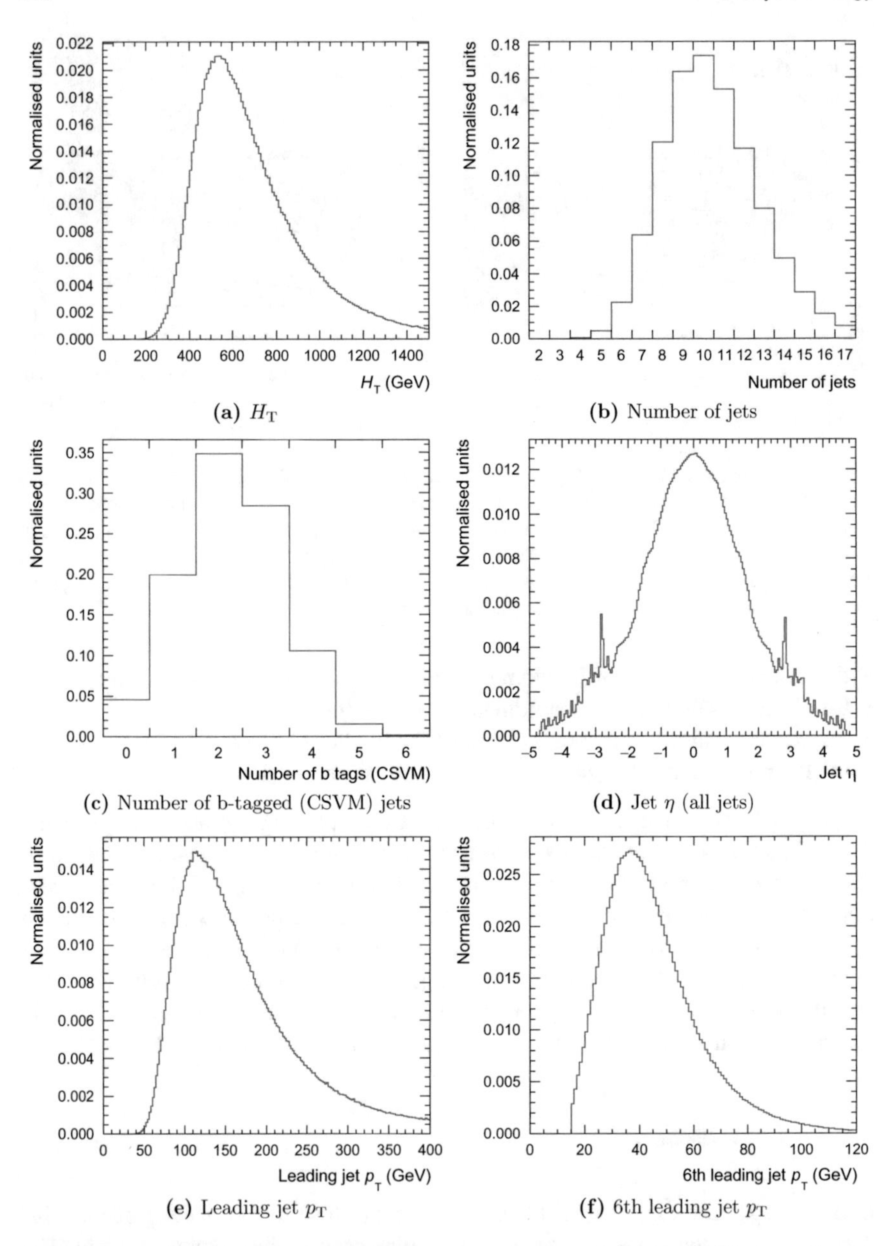

(a) H_{T} **(b)** Number of jets

(c) Number of b-tagged (CSVM) jets **(d)** Jet η (all jets)

(e) Leading jet p_{T} **(f)** 6th leading jet p_{T}

Fig. 6.8 Reconstruction-level variables in simulated fully hadronic $\mathrm{t\bar{t}H}$ ($\mathrm{H} \rightarrow \mathrm{b\bar{b}}$) events. There is no selection applied to these events, however a selection on the jets of $p_{\mathrm{T}} > 15\,\mathrm{GeV}$ and $|\eta| < 4.7$ is applied, and the H_{T} is calculated from these jets. The peaks appearing around $|\eta| = 2.9$ in (**d**) are likely caused by double counting of calorimeter-only jets around the transition zone between ECAL/HCAL and HF

- H_T greater than 300 GeV
- At least 6 jets
- At least 1 b-tagged jet
- Jets with $|\eta|$ less than 4
- Leading jet p_T greater than 60 GeV
- 6th leading jet p_T greater than 20 GeV

The above selection criteria were used to set the initial trigger thresholds during trigger development. Of course, the trigger rate at these thresholds would have been far too high, and thus all thresholds had to be immediately tightened. The final trigger thresholds are listed in Table 4.1, while the following preselection criteria are implemented:

- All jets with $p_T > 30$ GeV and $|\eta| < 2.4$
- H_T greater than 500 GeV
- At least 6 jets with $p_T > 40$ GeV
- At least 2 b-tagged jets

The η restriction on the jets ensures that they have tracks in the tracker, which results in better quality jets, but is also needed for b-tagging. The efficiencies of the trigger and each of these selection criteria, as well as various combined and relative efficiencies in fully hadronic $t\bar{t}H$ ($H \rightarrow b\bar{b}$) simulated events are listed in Table 6.4.

In addition to the selection criteria listed in Table 6.4, a veto on loose muons and electrons, respectively defined in Sects. 4.3.3 and 4.3.4, is applied to ensure that there is no overlap with the established leptonic $t\bar{t}H$ ($H \rightarrow b\bar{b}$) search. As previously mentioned, the efficiency of the lepton veto is very high and has little impact on the analysis selection. For the remainder of this thesis, *preselection* is used to refer to the certified data, passing the signal triggers, after the event cleaning, lepton veto and offline preselection. The same selection requirements are made for all MC simulated events, apart from the data certification.

To ensure a good understanding of the background processes contributing to the preselected events in data, the distributions of various reconstruction-level quantities have been compared in data and simulation. A selection of these distributions is shown in Fig. 6.9, which compares data to the simulated backgrounds, and also

Table 6.4 Efficiency in fully hadronic $t\bar{t}H$ ($H \rightarrow b\bar{b}$) simulated events of the trigger and preselection criteria, individually and combined. For the trigger, jets must have $|\eta| < 2.6$, while for the offline preselection they must have $|\eta| < 2.4$

Requirement	Efficiency (%)
Trigger	63.2
$H_T > 500$ GeV	54.8
$\geq$6 jets with $p_T > 40$ GeV	45.3
$\geq$2 b-tagged jets	59.5
Preselection	27.4
Trigger and preselection	27.1
Trigger w.r.t. preselection	99.2

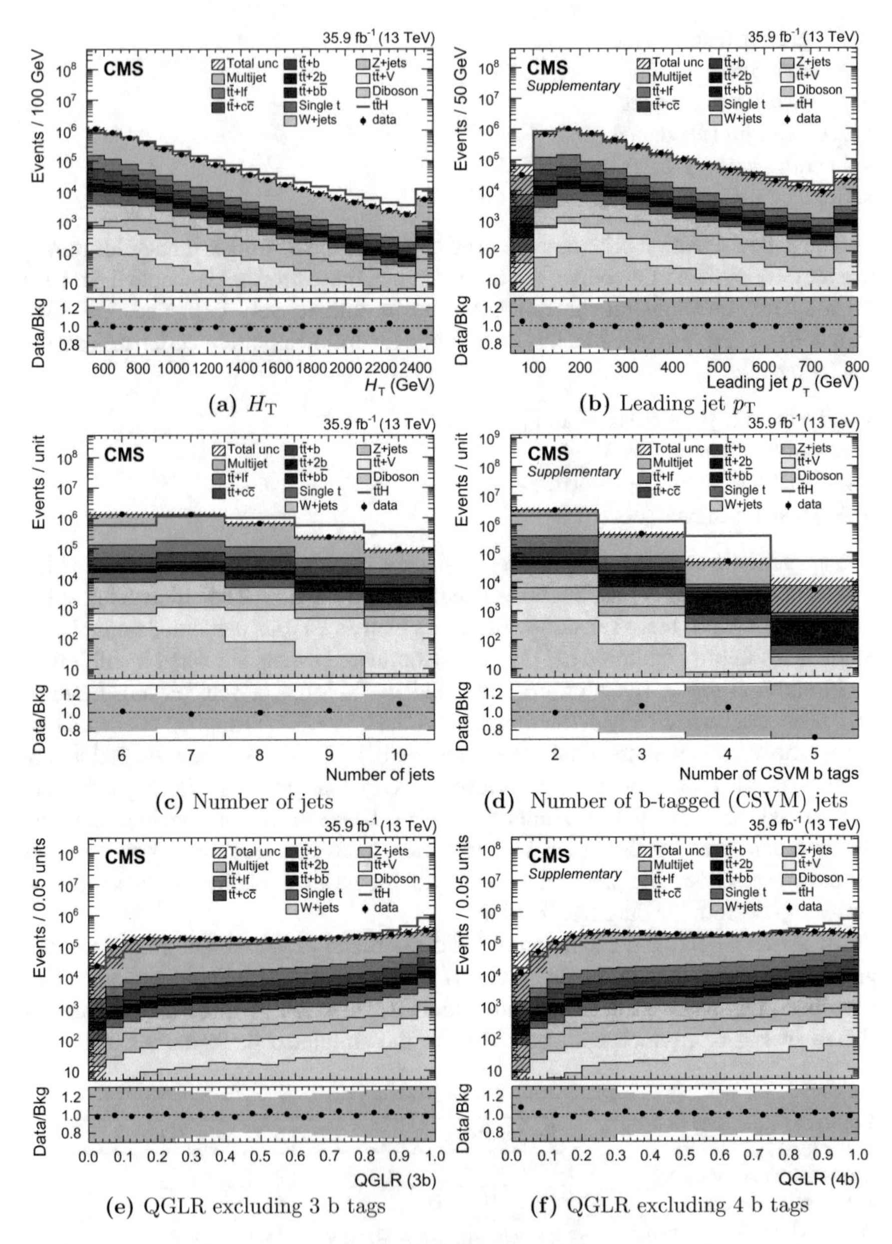

(a) H_{T}

(b) Leading jet p_{T}

(c) Number of jets

(d) Number of b-tagged (CSVM) jets

(e) QGLR excluding 3 b tags

(f) QGLR excluding 4 b tags

Fig. 6.9 Distributions of reconstruction-level variables in data (black points) and in simulation (stacked histograms) after the preselection. The simulated backgrounds are first scaled to the luminosity of the data, and then the simulated QCD multijet background is rescaled to match the yield in data. The contribution from the $\mathrm{t\bar{t}H}$ signal (blue line) is scaled to the total background yield (equivalent to the yield in data) to enhance readability. The striped error bands reflect the total statistical and systematic uncertainties on the backgrounds. The last bin includes event overflows. The ratios of data to background are given below the main panels, with the error bands reflecting the total uncertainties shows the signal distribution scaled to the total data yield to give an idea of the signal-background separation properties of each variable. The QCD multijet background has been estimated from MC simulation, but due to its large cross section uncertainty, it is scaled to fill the gap in yield between other simulated backgrounds and the data. The systematic uncertainties discussed in Sect. 6.6 are included in the total uncertainties shown in the preselection distributions of Fig. 6.9.

6.3.2 Final Selection

After the preselection, the signal is still overwhelmed by background events, and a further selection must be made in order to increase the signal purity. Based on the discriminating power of jet and b-tag multiplicity observed in Fig. 6.9c and d, a categorisation is made using these two variables. Specifically, six categories are formed for the signal region with at least 7 jets and at least 2 b tags, while three categories with exactly 2 b tags are used for a control region from which to estimate the QCD multijet background, as described in Sect. 6.4.2. The signal and control categories are indicated in Table 6.5.

The decision to not use fewer than 7 jets or less than 3 b-jets for the signal region is driven by the overwhelming QCD multijet background and very low signal contribution at lower multiplicities. The 7 jet, 3 b-jet category already has a low signal and large multijet contribution and serves to constrain uncertainties much more than provide sensitivity to the signal.

In addition to the jet and b-tag requirements, to reject events that are unlikely to include a W boson from top quark decays, a cutoff is placed on the dijet invariant mass. All untagged jets are considered in the calculation, and the invariant mass of the pair closest to m_{W} ($m_{\mathrm{qq'}}$) is chosen as the W mass for the event. The following requirements are applied, which have shown to increase the discriminating power of the matrix element method (MEM) discriminant compared to not applying any cutoff on the dijet invariant mass:

- 7 jets: $60 < m_{\mathrm{qq'}} < 100\,\mathrm{GeV}$
- 8 jets: $60 < m_{\mathrm{qq'}} < 100\,\mathrm{GeV}$
- ≥9 jets: $70 < m_{\mathrm{qq'}} < 92\,\mathrm{GeV}$

Table 6.5 Event categories considered in the analysis

Number of b tags	Number of jets		
	7	8	≥9
≥4	Signal	Signal	Signal
3	Signal	Signal	Signal
2	Control	Control	Control

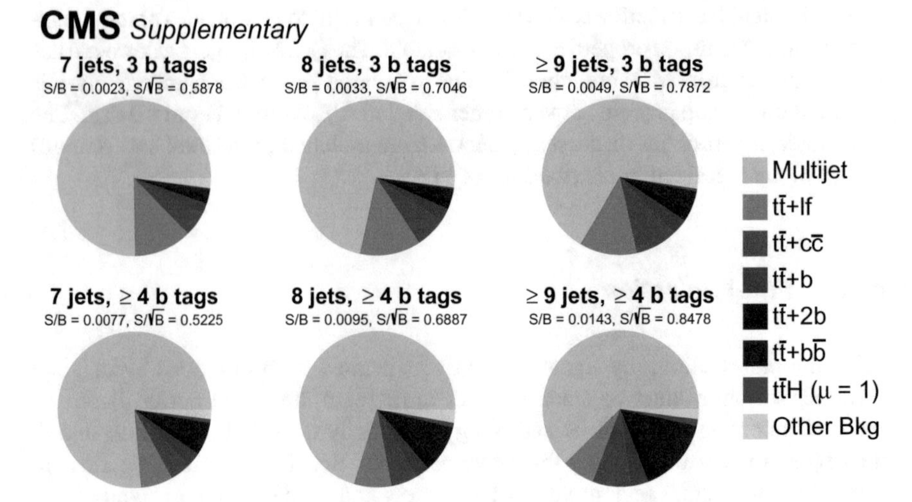

Fig. 6.10 Expected fraction of signal and background processes contributing to the analysis categories. Figure produced by [Korbinian Schweiger, UZH]

As described in Sect. 4.3.7, a selection requirement is also placed on the QGLR calculated based on the number of b-tagged jets in the event (Figs. 6.9e and f). The requirement of QGLR > 0.5 is applied, which has a signal efficiency of around 76%, and an efficiency in data of around 56%. This QGLR selection not only helps increase the signal purity, but it also allows for a validation region of the QCD multijet estimation method, as described in Sect. 6.4.2.

The expected contribution of the signal and background processes to each signal region category, after the W mass requirement and the QGLR selection, is shown in Fig. 6.10. From this point forward, all references to an *analysis category* or *signal region category* include the aforementioned selections on the W mass and the QGLR.

6.4 Background Estimation

The main background processes in this search stem from QCD multijet and $\mathrm{t\bar{t}}$ production associated with additional light-flavour, charm, or bottom quarks ($\mathrm{t\bar{t}}+\mathrm{jets}$). The background from $\mathrm{t\bar{t}}+\mathrm{jets}$ as well as other minor backgrounds (single top quark, $\mathrm{V}+\mathrm{jets}$, $\mathrm{t\bar{t}}+\mathrm{V}$, and diboson events) are estimated through MC simulation, while a data-driven technique has been developed to model the background from QCD multijet events. The validation of $\mathrm{t\bar{t}}+\mathrm{jets}$ (and minor backgrounds), and the derivation and validation of the QCD multijet background are discussed in the remainder of this section.

6.4.1 $t\bar{t}$ + jets Background

In order to validate that the MC simulation of $\mathrm{t\bar{t}}$ events is a good representation of the data, a validation region enriched in $\mathrm{t\bar{t}}$ events has been defined. For this purpose, the semi-leptonic top-quark-decay channel to muons is used, where the background from QCD multijet events is negligible.

Events in the validation region are selected by a single-muon trigger with a p_{T} requirement of more than 24 GeV. The offline reconstruction then requires a tight muon (see Sect. 4.3.3) with $p_{\mathrm{T}} > 26$ GeV, and at least four reconstructed jets with $p_{\mathrm{T}} > 30$ GeV out of which at least two are b tagged (CSVM). In addition to the reweighting discussed in Sect. 6.2 (pileup, top p_{T}, b-tagging, and QGL), the simulated samples are corrected using the scale factors for the muon trigger and identification, as derived centrally by the CMS muon group.

A few comparisons between data and simulated events in the single-muon validation region have already been shown in Figs. 6.4b, d and f, while some additional variables are compared in Fig. 6.11. A reasonable agreement between simulation and data is observed in all kinematic distributions, thus validating the use of MC simulation for the $\mathrm{t\bar{t}}$ + jets and other minor background processes. The slight discrepancy at high jet multiplicity is a known issue in $\mathrm{t\bar{t}}$ simulation and is accounted for with systematic uncertainties.

In addition to validating standard kinematic variables, the single-muon validation region is also used to validate the QGLR, which is defined analogously to the signal region. The subset of events with at least 5 (6) jets in the validation region is used to cross check the QGLR in the signal region for 3b (4b) events. The distribution of the QGLR in the validation region predicted by simulation agrees well with data, as shown in Fig. 6.12, which demonstrates the validity of the QGLR for the use in this analysis.

6.4.2 QCD Multijet Background

The QCD multijet background is derived from data by using a control region with low b-tag multiplicity to estimate the contribution from QCD multijet events in the signal region. The control region is enriched in QCD multijet events, and the remaining contribution from other backgrounds (mainly $\mathrm{t\bar{t}}$ + jets) is subtracted using simulation.

The control region is defined by events with two CSVM and one or more additional CSVL jets. In addition, a validation region, in which to test the multijet estimation method, is defined by events with QGLR < 0.5. This provides four orthogonal regions, summarised in Table 6.6, from which the multijet background estimate can be obtained and validated. The use of the validation region relies on the fact that the QGLR and the number of additional CSVL jets are uncorrelated by construction, which has also been verified in simulation and data. The four orthogonal regions are

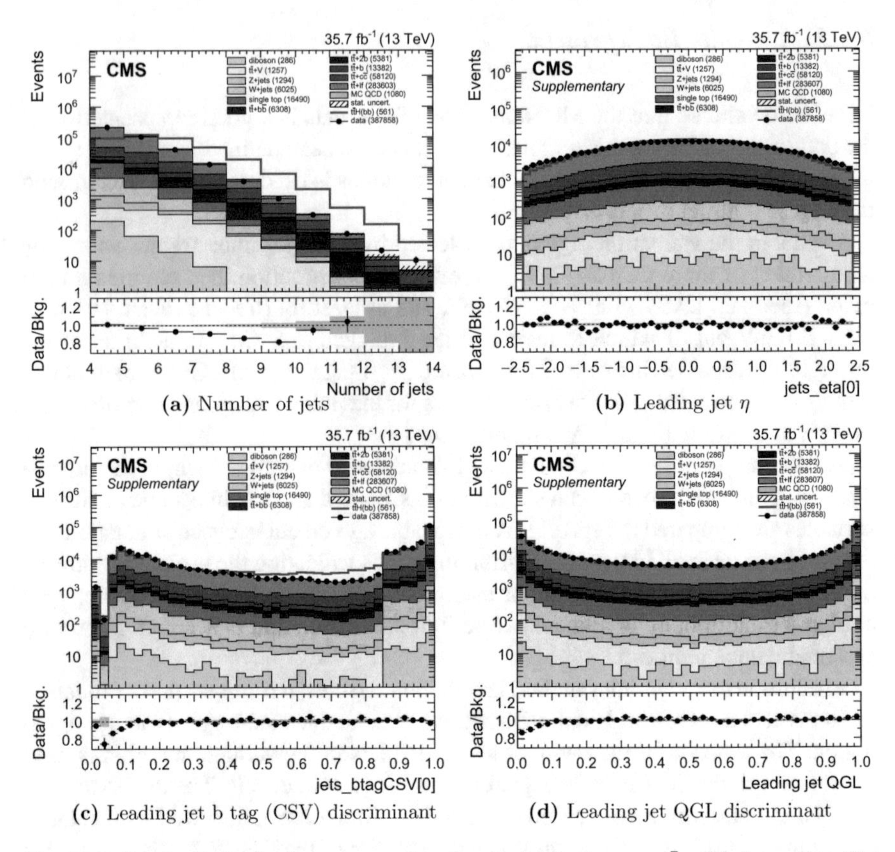

(a) Number of jets (b) Leading jet η

(c) Leading jet b tag (CSV) discriminant (d) Leading jet QGL discriminant

Fig. 6.11 Simulated distributions compared to data in the single-muon $t\bar{t}$ validation region. The simulated backgrounds are scaled to the luminosity of the data. The uncertainty bands include statistical uncertainties only

used independently in each of the six analysis categories defined in Table 6.5. For a given variable, the distribution in multijet events in the signal region of each category is estimated from the data in the control region, after subtracting $t\bar{t}$ + jets and other minor background processes.

Since the kinematic properties of jets differ in the control and signal regions because of different heavy-flavour composition, corrections as a function of jet p_{T}, η, and the minimum distance from the first two b-tagged jets (ΔR_{min}) are applied to the one or two CSVL jets in the control regions. The correction method is described later in this section and is intended to reweight the kinematic distributions of CSVL jets to match those of CSVM jets. The corrected multijet distribution in the control region is then scaled to provide an estimate of the distribution in the signal region. The exact scaling is determined in the final fitting procedure (see Chap. 7), where the multijet yield in each category is left floating. The initial value of the multijet normalisation is unimportant and set to the yield in data less the yield of simulated

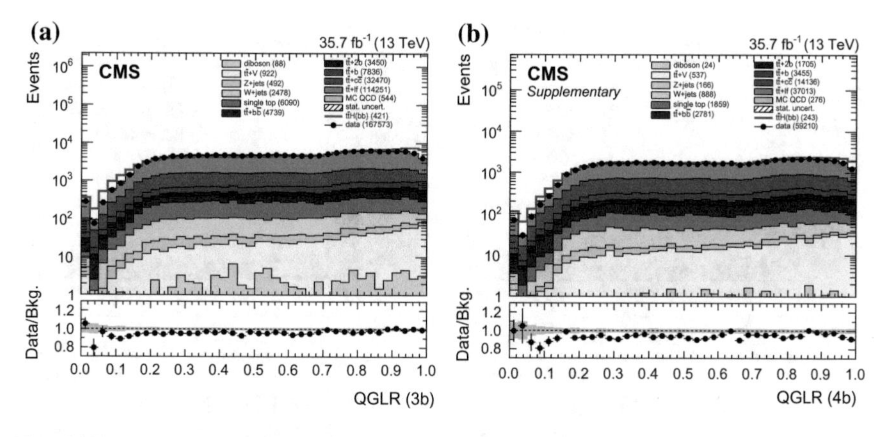

Fig. 6.12 Comparison of the QGLR distribution in data and simulation in the single-muon $t\bar{t}$ validation region. (**a**) Events with at least 5 jets and excluding the first 3 b-tagged jets. (**b**) Events with at least 6 jets and excluding the first 4 b-tagged jets. The simulated backgrounds are scaled to the luminosity of the data. Uncertainty bands include statistical uncertainties only

Table 6.6 Definition and description of the four orthogonal regions in the analysis

	$N_{\mathrm{CSVM}} = 2$ $N_{\mathrm{CSVL}} \geq 3$	$N_{\mathrm{CSVM}} \geq 3$
QGLR > 0.5	Control region (CR) (to extract distribution)	Signal region (SR) (final analysis)
QGLR < 0.5	Validation control region (CR2) (to validate distribution)	Validation region (VR) (comparison with data)

events in the signal region, where the simulated events include all other background and signal processes. The ratio of the signal region yield to that in the control region is of the order of 0.4 for the 3 b-tag categories and 0.1 for the 4 b-tag categories.

A consistency check of the procedure used to estimate the multijet background is performed in simulation, and agreement is observed within the statistical uncertainties. Since the simulated multijet events are quite low in terms of statistics, the power of this test is limited. A better validation of the method is performed in data using events with QGLR < 0.5, by applying the same procedure used to estimate the multijet background in the signal region. The distribution of the final MEM discriminant in the six categories of the validation region, VR, in data, together with the data-driven multijet estimate, and other simulated background contributions is shown in Fig. 6.13.

In addition to the validation of the MEM distribution, the multijet estimation method has been validated with other kinematic variables. During this validation, a slight discrepancy was observed in events with low H_{T} in the 7j, 3b category and all 4b categories. In light of this, two uncorrelated systematic uncertainties are applied, as discussed in Sect. 6.6. The validation region distributions of a selection of kinematic

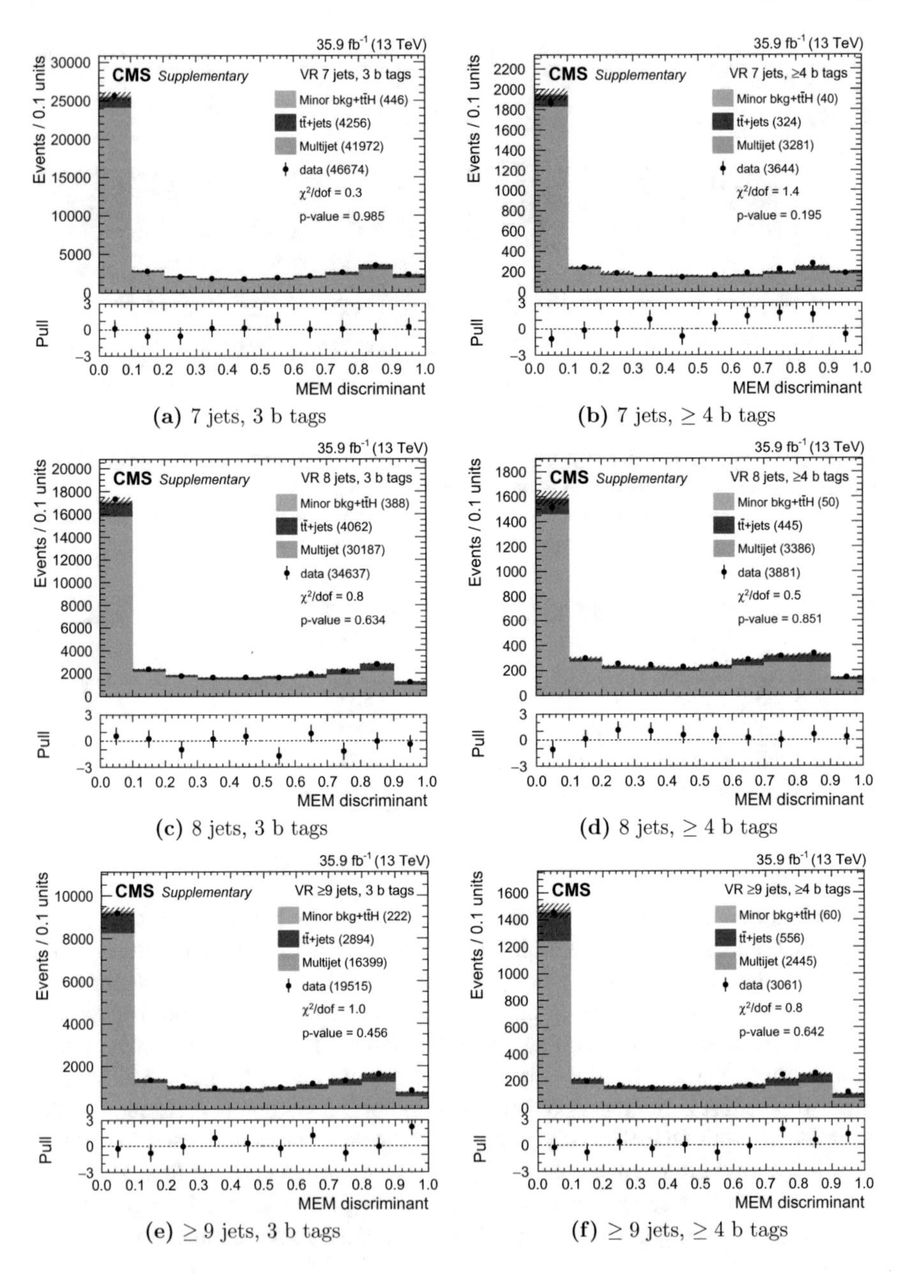

(a) 7 jets, 3 b tags

(b) 7 jets, ≥ 4 b tags

(c) 8 jets, 3 b tags

(d) 8 jets, ≥ 4 b tags

(e) ≥ 9 jets, 3 b tags

(f) ≥ 9 jets, ≥ 4 b tags

Fig. 6.13 Distributions in the MEM discriminant in data, simulated backgrounds, and the estimated multijet background in the six categories of the validation region. The level of agreement between data and estimation is expressed in terms of a χ^2 divided by the number of degrees of freedom (dof), and the corresponding p-values are also shown. The differences between data and the total estimates divided by the total statistical and systematic uncertainties in the data and estimates (pulls) are given below the main panels. The numbers in parenthesis in the legend represent the total yields for the corresponding entries

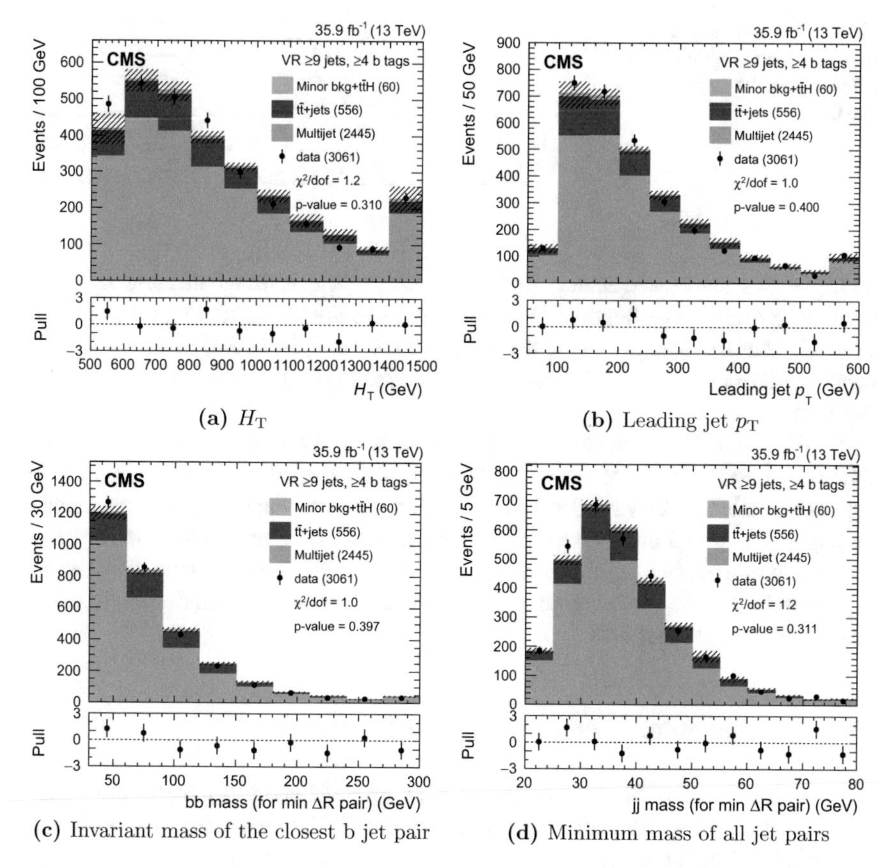

(a) H_{T} (b) Leading jet p_{T}

(c) Invariant mass of the closest b jet pair (d) Minimum mass of all jet pairs

Fig. 6.14 Various distributions in data, simulated backgrounds, and the estimated multijet background in the ≥9j, ≥4b category of the validation region. The level of agreement between data and estimation is expressed in terms of a χ^2 divided by the number of degrees of freedom (dof), and the corresponding p-values are also shown. The differences between data and the total estimates divided by the total statistical and systematic uncertainties in the data and estimates (pulls) are given below the main panels. The numbers in parenthesis in the legend represent the total yields for the corresponding entries

variables for the most sensitive event category, ≥9j, ≥4b, are shown in Fig. 6.14, including all systematic uncertainties described in Sect. 6.6 except for the multijet normalisation. All discrepancies are accounted for by the systematic uncertainties applied to the multijet and other backgrounds, as can be seen by the relatively low χ^2 in the figures.

To verify that the good performance of the method demonstrated in the gluon-jet enriched validation region (QGLR < 0.5) also holds in the quark-jet enriched signal region (QGLR > 0.5), another control region (side band) has been investigated. Specifically, the b tagging criteria were changed by selecting jets that fulfil an intermediate b tagging requirement (CSVML), which is then used to form jet and b jet

Table 6.7 Definition of the four orthogonal regions derived from the original CR2 and CR

	$N_{\text{CSVM}} = 2$	
	$N_{\text{CSVML}} = 2$ $N_{\text{CSVL}} \geq 3$	$N_{\text{CSVML}} \geq 3$
QGLR > 0.5	CRx	SRx
QGLR < 0.5	CR2x	VRx

multiplicity categories in analogy with the signal region, as shown in Table 6.7. Since these categories are orthogonal to the categories in the signal region, they are used to verify that the background estimation is valid for QGLR > 0.5. The results of these validations for the MEM discriminant in the high-QGLR region are shown in Fig. 6.15.

In addition, a further control region has been investigated in which the division in QGLR is made from 0.0 to 0.3 and 0.3 to 0.5, while the b-tagging requirements are kept as in Table 6.6. This validates that there is no obvious dependency of the method on the QGLR range considered. In all validation regions, the multijet background estimation reproduces, within the assigned uncertainties, the kinematic distributions measured in data. In the remainder of this thesis, all multijet estimates are based on data unless stated otherwise.

Kinematic Correction to Loose B-Tagged Jets

Jets passing the CSVM tag are observed to have different kinematic properties, specifically p_{T}, η, and the minimum distance from the first two b-tagged jets, ΔR_{min}, than those failing the b tag.

Based on these kinematic differences, a correction has been developed to apply to CSVL jets in data and simulation to alter their kinematic distributions to match those of CSVM jets. Specifically, the distribution of the ratio of CSVM jets to CSVL jets in p_{T}–η–ΔR_{min} space is calculated in events passing the preselection,[2] excluding the first two jets ordered according to CSV output, and then a 1-dimensional function is fitted to the projection of this distribution on each of the p_{T}, η, and ΔR_{min} axes. The forms of the functions are as follows:

$$p_{\text{T}} : f(x) = p_0 + \text{erf}(p_1(x - p_2)) \cdot (p_3 + p_4 x), \tag{6.4}$$

$$\eta : g(x) = \sum_{i=0}^{16} p_i x^i, \tag{6.5}$$

[2]Although the correction is derived in the preselection region, i.e. inclusive in QGLR, it is used for both QGLR > 0.5 and QGLR < 0.5, in the search and validation region, respectively. The dependency of the correction on QGLR has been investigated and found to be consistent within statistical uncertainties, with an ultimate negligible effect on the multijet MEM distribution.

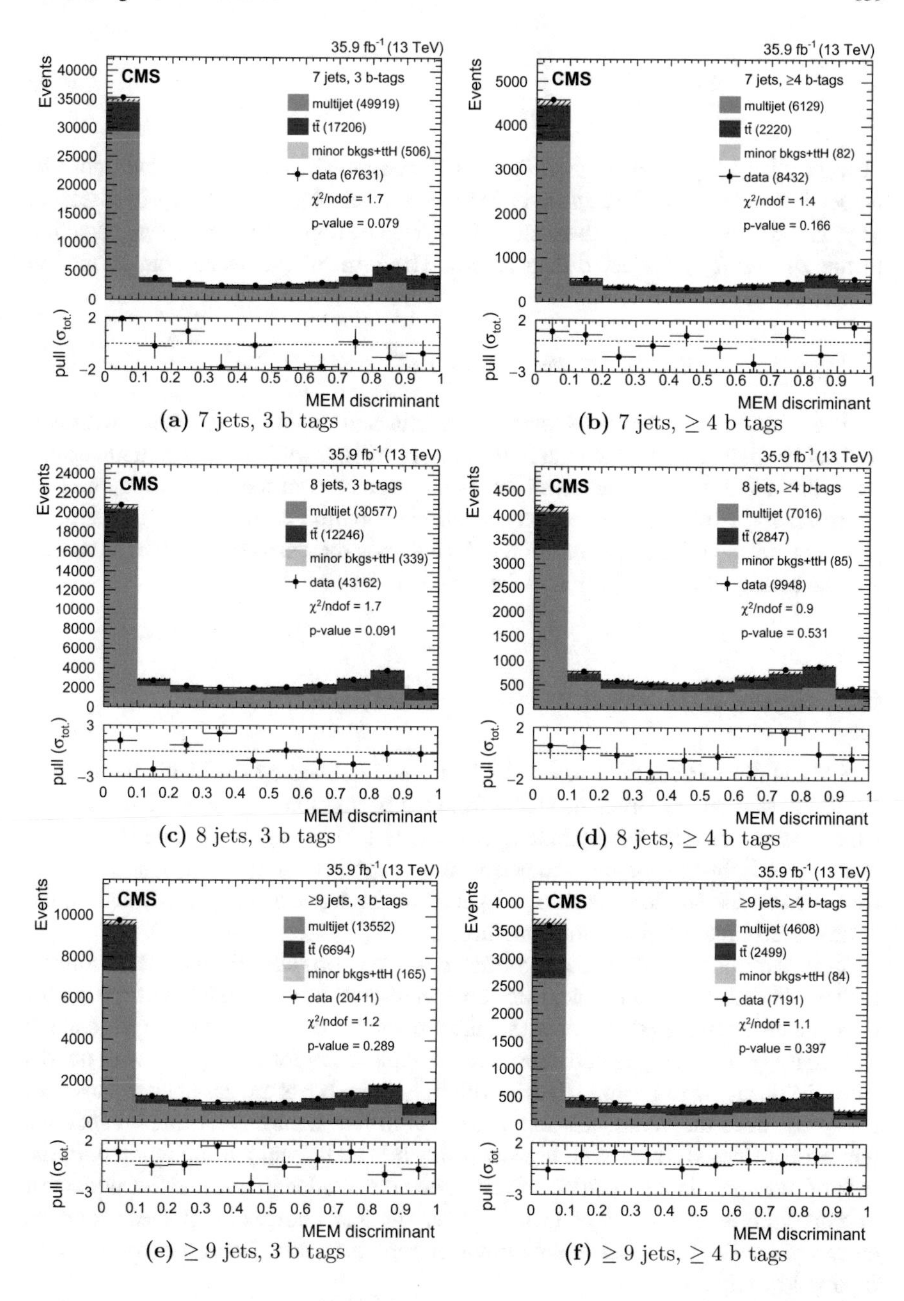

(a) 7 jets, 3 b tags
(b) 7 jets, ≥ 4 b tags
(c) 8 jets, 3 b tags
(d) 8 jets, ≥ 4 b tags
(e) ≥ 9 jets, 3 b tags
(f) ≥ 9 jets, ≥ 4 b tags

Fig. 6.15 Distributions in the MEM discriminant in data, simulated backgrounds, and the estimated multijet background in the six categories of the side-band signal region, SRx. The level of agreement between data and estimation is expressed in terms of a χ^2 divided by the number of degrees of freedom (dof), and the corresponding p-values are also shown. The differences between data and the total estimates divided by the total statistical uncertainties in the data and estimates (pulls) are given below the main panels. The numbers in parenthesis in the legend represent the yields in the visible range for the corresponding entries

$$\Delta R_{\min} : h(x) = \sum_{i=0}^{6} p_i x^i. \tag{6.6}$$

The product, $f(p_T) \cdot g(\eta) \cdot h(\Delta R_{\min})$ is then applied to the one or two loose b-tagged jets in an event. Since a slight dependency between η and $\Delta R_{\min}$ is observed, a systematic uncertainty on the method is derived by applying a second η-correction to the p_T-η-$\Delta R_{\min}$ corrected distribution. The form of the second eta correction function is as follows:

$$\eta : g_2(x) = p_0 + p_1 x + p_2 x^2 + p_3 x^3 + p_4 x^4. \tag{6.7}$$

The correction is derived separately in data and simulated $\mathrm{t\bar{t}}$ events, with that derived from data applied to data in the control region, while the correction derived from simulated $\mathrm{t\bar{t}}$ is applied to all simulated processes in the control region. This approach is justified given the very small contribution of the minor background processes in the control region. Figure 6.16 shows the derivation of the correction and the results of its application in data and simulation.

6.5 Signal Extraction

The MEM discriminant discussed in Chap. 5 and defined by the likelihood ratio in Eq. (5.72) is used to extract the signal from the background. Although the discriminant is constructed to discriminate against the $\mathrm{t\bar{t}} + \mathrm{b\bar{b}}$ background, it performs well against $\mathrm{t\bar{t}}$ + light-flavour jets, and performs best against multijet events, and is therefore used as the single discriminant against all backgrounds. By construction, the MEM discriminant satisfies the condition $0 \leq \mathcal{P}_{\mathrm{S/B}} \leq 1$.

In each event, the three or four jets that most likely originate from b jets (according to their CSV discriminant values) are considered explicitly as candidates for b quarks from the decay of the Higgs boson and the top quarks, whereas untagged jets and the fifth or more b-tagged jets are considered as candidates for the light-flavour quarks from the decay of W bosons. Events with only three b jets are assumed to have lost a b quark from the decay of a top quark. Up to five light-flavour quark candidate jets are considered (those with highest p_T), while additional jets are ignored. In the case of five light-flavour quark candidates, one is excluded in turn and the number of permutations is increased by a factor of five. The final choice of hypothesis for each category, considering discrimination power and computing performance, has been reported in Table 5.6.

The final MEM discriminants in data, the different background processes, and the signal, are shown in Fig. 6.17 for each analysis category before the fit to data. The event yields expected for the signal and the different background processes for an integrated luminosity of 35.9 fb^{-1} after applying the pileup reweighting, b-tagging

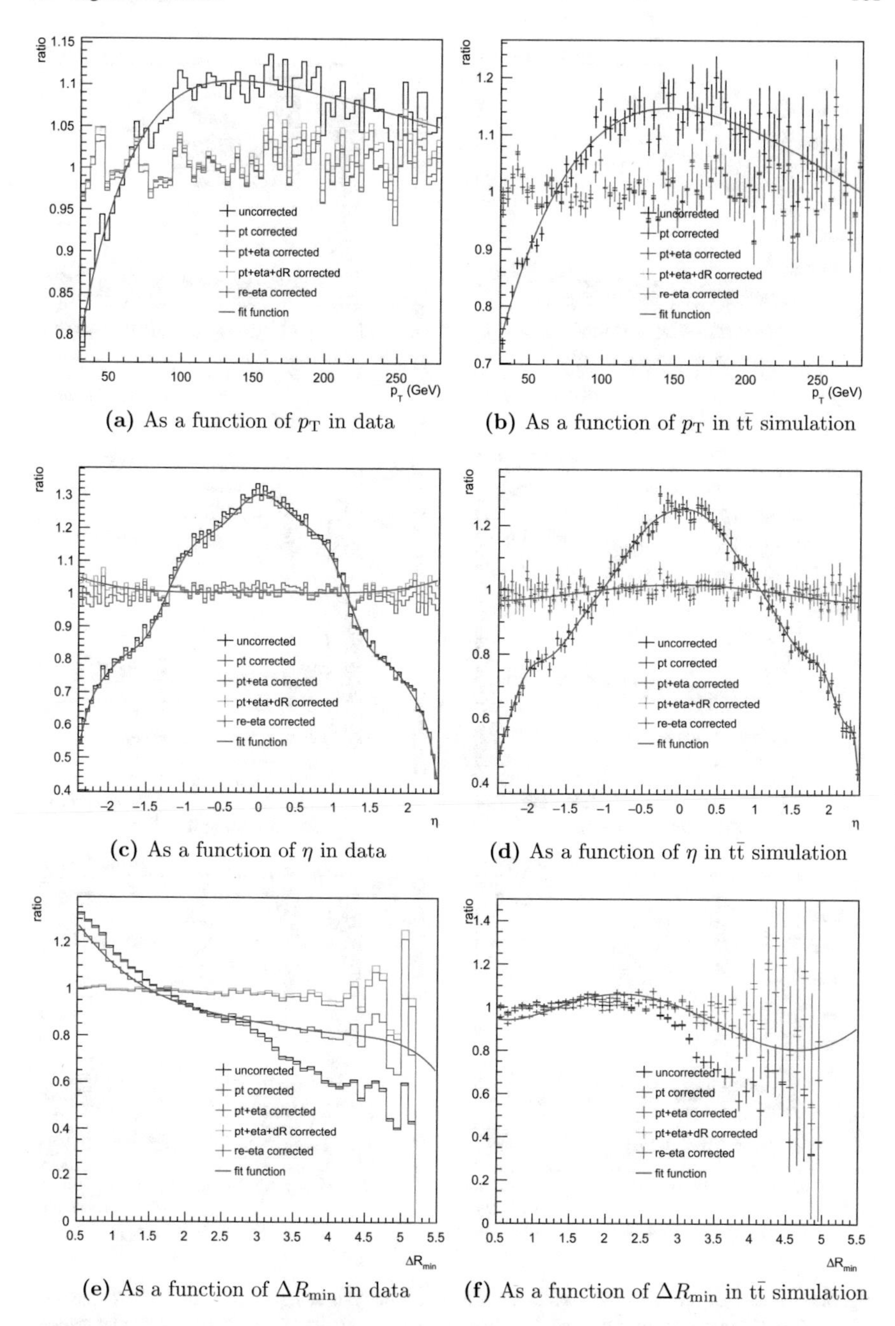

(a) As a function of p_T in data

(b) As a function of p_T in $t\bar{t}$ simulation

(c) As a function of η in data

(d) As a function of η in $t\bar{t}$ simulation

(e) As a function of ΔR_{min} in data

(f) As a function of ΔR_{min} in $t\bar{t}$ simulation

Fig. 6.16 Distribution of the ratio of the number of CSVM jets to the number of CSVL jets as a function of p_T, η and ΔR_{min} after the preselection and excluding the first two jets by CSV, before applying the correction (black), after the p_T correction (blue), after the p_T and η correction (red), after the p_T, η and ΔR_{min} correction (green) and after the η re-correction (magenta). The fitted functions, $f(p_T)$, $g(\eta)$, $h(\Delta R_{min})$ and $g_2(\eta)$ are shown as thick red lines

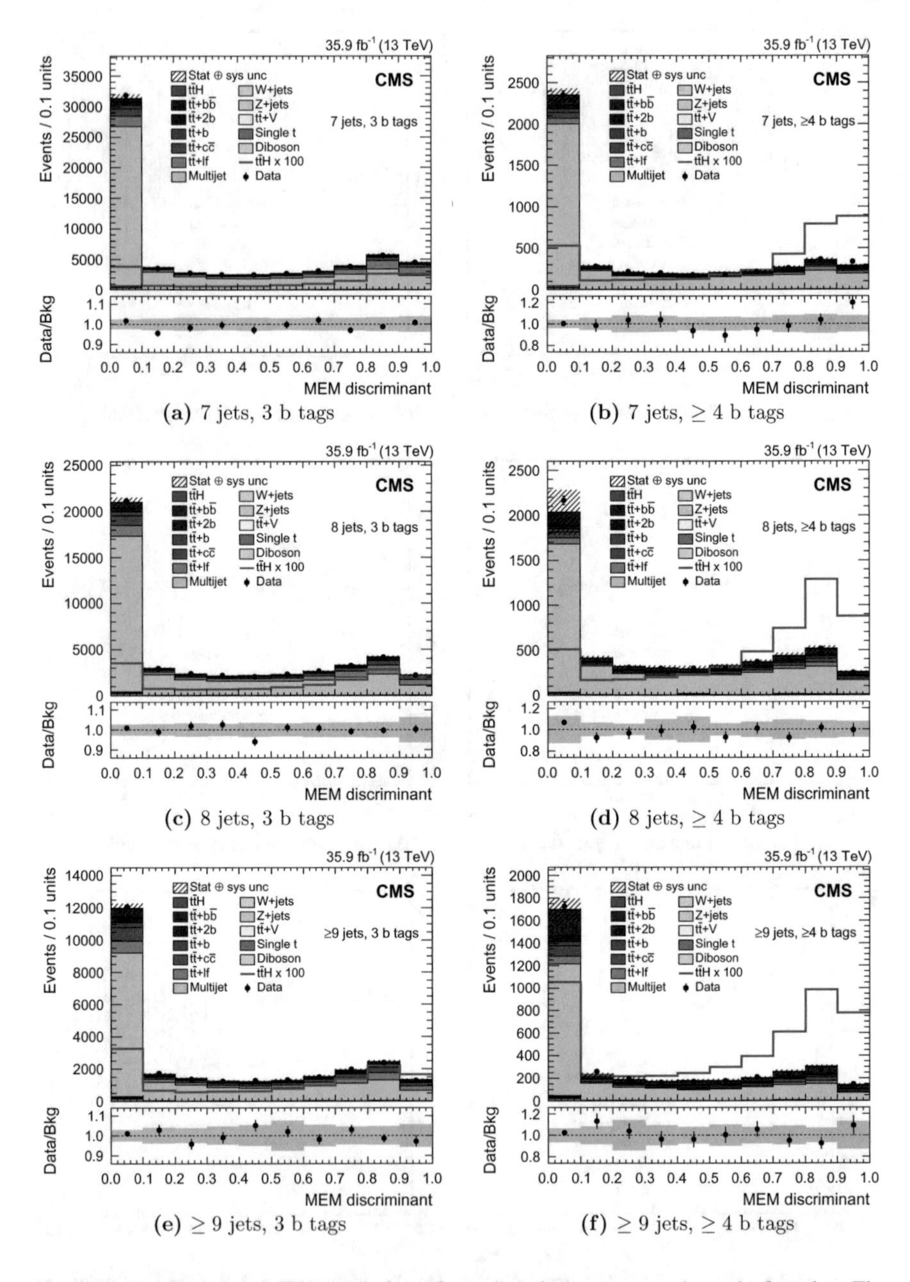

(a) 7 jets, 3 b tags

(b) 7 jets, $\geq$ 4 b tags

(c) 8 jets, 3 b tags

(d) 8 jets, $\geq$ 4 b tags

(e) $\geq$ 9 jets, 3 b tags

(f) $\geq$ 9 jets, $\geq$ 4 b tags

Fig. 6.17 Distributions in MEM discriminant for each analysis category prior to the fit to data. The expected contributions from signal and background processes (filled histograms) are shown stacked. The expected signal distributions (lines) for a Higgs boson mass of $m_{\mathrm{H}} = 125$ GeV are multiplied by a factor of 100 and superimposed on the data. Each background contribution is initially normalised to an integrated luminosity of 35.9 fb^{-1}, while the multijet contribution in each category is scaled to match the yield in data. The distributions observed in data (data points) are also shown. The ratios of data to background are given below the main panel

Table 6.8 Expected number of $t\bar{t}H$ signal and background events, and the observed event yields for the six analysis categories, prior to the fit to data. The yield of the multijet background is scaled such that the total background plus signal yield matches the yield in data. The quoted uncertainties contain all pre-fit uncertainties described in Sect. 6.6 added in quadrature, considering all correlations among processes. The signal (S) and total background (B) ratios are also shown

Process	7j, 3b	8j, 3b	≥9j, 3b	7j, ≥4b	8j, ≥4b	≥9j, ≥4b
Multijet	46 608 ± 4 606	32 220 ± 3 487	17 300 ± 2 435	3 528 ± 438	3 824 ± 620	2 211 ± 508
$t\bar{t}$ + lf	7 587 ± 2 279	5 419 ± 1 567	2 915 ± 854	250 ± 177	267 ± 344	191 ± 155
$t\bar{t} + c\bar{c}$	3 631 ± 2 016	3 275 ± 1 781	2 426 ± 1 338	196 ± 172	285 ± 250	269 ± 249
$t\bar{t}$ + b	1 424 ± 679	1 184 ± 655	849 ± 427	131 ± 86	146 ± 110	123 ± 83
$t\bar{t}$ + 2b	989 ± 530	818 ± 431	639 ± 332	90 ± 68	111 ± 85	103 ± 61
$t\bar{t} + b\bar{b}$	1 194 ± 574	1 373 ± 614	1 284 ± 610	278 ± 147	485 ± 245	534 ± 272
Single t	755 ± 223	514 ± 159	288 ± 93	43 ± 22	67 ± 70	37 ± 21
W+jets	380 ± 189	195 ± 105	135 ± 290	16 ± 20	19 ± 125	9 ± 15
Z+jets	78 ± 26	86 ± 32	61 ± 24	6 ± 5	9 ± 6	11 ± 6
$t\bar{t}$ +V	113 ± 24	120 ± 31	111 ± 36	13 ± 8	23 ± 37	28 ± 16
Diboson	14 ± 7	6 ± 5	2 ± 1	0.7 ± 0.7	0.9 ± 0.8	0 ± 6
Total bkg	62 773 ± 1 801	45 209 ± 1 513	26 009 ± 1 081	4 553 ± 245	5 237 ± 506	3 516 ± 296
$t\bar{t}H$	147 ± 31	150 ± 27	127 ± 25	35 ± 9	50 ± 16	50 ± 15
Data	62 920	45 359	26 136	4 588	5 287	3 566
S/B	0.002	0.003	0.005	0.008	0.010	0.014
$S/\sqrt{B}$	0.59	0.70	0.79	0.52	0.69	0.85

scale factors, top-p_T reweighting, QGL reweighting and trigger scale factors, and the yields observed in data are listed in Table 6.8, and also shown in Fig. 6.18, for each category.

6.6 Systematic Uncertainties

There are several sources of systematic uncertainty that can affect the expected amount of signal and background in each bin of the MEM discriminant. Each independent source is associated with a *nuisance parameter*[3] that modifies the likelihood in the final fit described in Sect. 7.1, and can either affect the yield from a process (rate uncertainty), or the distribution in the MEM discriminant (shape uncertainty), or both. In the latter case, the effects on the rate and shape are treated simultaneously and are considered completely correlated. Each individual source of systematic uncertainty is independent of other sources, and its effect on signal and background is 100% correlated across the processes to which it applies. A description of the systematic uncertainties considered for this analysis is provided in the following. Unless otherwise noted, each systematic uncertainty applies equally to the signal

[3]A nuisance parameter is a parameter which is not of immediate interest but can influence the statistical model, thus affecting the parameters which are of interest.

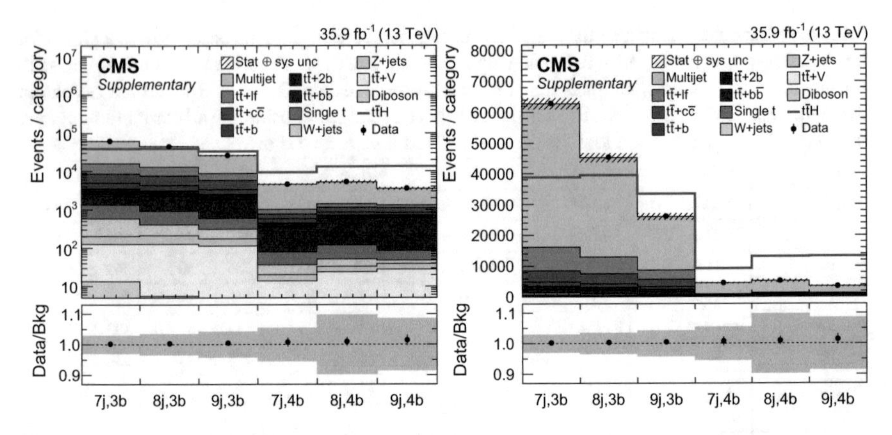

Fig. 6.18 Predicted (histograms) and observed (data points) event yields in each analysis category prior to the fit to data, corresponding to an integrated luminosity of 35.9 fb^{-1}. The expected contributions from different background processes (filled histograms) are stacked, showing the total pre-fit uncertainty (striped error bands), and the expected signal distribution (line) for a Higgs boson mass of $m_{\text{H}} = 125$ GeV is scaled to the total background yield for ease of readability. The ratios of data to background are given below the main panels, with the full uncertainties. The yields are shown with a logarithmic scale (left) and linear scale (right)

and all simulated background processes. The data-driven QCD multijet estimate is only impacted by the systematic uncertainties affecting MC simulation through the subtraction of simulated processes from data in the control region.

Jet energy scale The impact of the uncertainty on the jet energy scale (JES) correction [25] is evaluated for each jet in the simulated events by changing the correction factors by their uncertainties, and propagating the effect to the MEM discriminant by recalculating all kinematic quantities. 25 independent sources contribute to the overall JES uncertainty, and therefore their impact is evaluated separately and they are treated as uncorrelated in the final fit. Since the analysis categories are defined in terms of jet multiplicity and kinematics, a change in JES can induce a migration of events between analysis categories, as well as in or out of the signal region. The fractional change in event yields induced by a one-standard-deviation shift in JES ranges between 3–11%, depending on the process and the category.

Jet energy resolution The uncertainty related to jet energy resolution (JER) is evaluated by increasing and decreasing the difference between the reconstructed-level and particle-level jet energy, according to the standard CMS prescription. It ranges between about 1 and 5% of the expected jet energy resolution, depending on the jet direction. The effect of the JER is accounted for in a similar way to the JES, by recalculating all kinematic quantities. The fractional change in event yields following the migration of events between analysis categories induced by a one-standard-deviation shift in JER ranges between 2–11%, which is again process and category dependent.

Integrated luminosity The uncertainty in the integrated luminosity affects the rate of all simulated processes. As mentioned in Sect. 6.1.1, it is estimated to be 2.5% [2].

Pileup reweighting As described in Sect. 6.2.1, the uncertainty in the distribution in the number of pileup interactions is evaluated by changing the minimum-bias cross section by 4.6% relative to its nominal value. The changes in the resulting weight factor are propagated to the MEM discriminant and treated as fully correlated among all simulated processes.

Top p_{T} reweighting As described in Sect. 6.2.2, the p_{T} of the generated top quark in simulated $\mathrm{t\bar{t}}$ events is reweighted according to the results of the differential $\mathrm{t\bar{t}}$ cross section measurements in Ref. [22]. The systematic uncertainty of this procedure is assessed by varying the reweighting function within its uncertainty (see the alternative functions in Fig. 6.3b). These two functions are used as "up" and "down" variations of the systematic uncertainty, which only affects simulated $\mathrm{t\bar{t}}$ events, and impacts both the rate and distribution in the MEM discriminant.

Trigger scale factors As described in Sect. 6.2.3, the uncertainties in trigger-scale factors are determined by the bin-by-bin uncertainty on the ratio of efficiency in data relative to simulation, and are approximately 1.5% on average, with some as high as 15%. The systematic variations are derived by assigning event weights of $\mathrm{SF} \pm \sigma_{\mathrm{SF}}$, and have a small impact on both the yield and distribution of the MEM discriminant.

B-tagging scale factors The scale factors applied to correct the CSV discriminant, described in Sect. 6.2.4, are affected by several components of systematic uncertainty, which are attributed to three main sources: JES, purity of heavy- or light-flavour jets in the control sample used to obtain the scale factors, and the statistical uncertainty of the event sample used in their extraction. A separate, large uncertainty is applied to charm-flavour jets owing to the lack of a reliable data-based calibration, namely the nominal scale factor is set to unity while its uncertainty is taken as twice that attributed to the heavy-flavour scale factor. Each component of these systematic b-tagging uncertainties is considered uncorrelated from the others, resulting in nine separate nuisance parameters in the final fit.

QGL reweighting A systematic uncertainty is assigned to the event based reweighting of each jet's QGL distribution discussed in Sect. 6.2.5. The uncertainty is taken as the full correction difference, i.e. the nominal is taken as the QGL reweighed event, the down variation is without QGL reweighting and the up variation is with twice the QGL reweighting minus one. Although the QGL reweighting does not affect the yield of a given process inclusive in QGL, the requirement of QGLR > 0.5 ensures that the yield in the signal region is impacted. This uncertainty therefore affects the rate and distribution of the MEM discriminant.

Process cross sections The expectation for MC based signal and background yields are derived from theoretical predictions of at least NLO accuracy. These normalisations are affected by uncertainties from QCD factorisation and renormalisation scales (QCD scale) and PDF uncertainties, which are summarised in Table 6.9 and considered rate uncertainties in the final fit. Where appropriate, the QCD scale uncertainties are treated as fully correlated for related processes, while the PDF uncertainties are considered fully correlated for all processes that share the same dominant initial state (i.e. gg, gq, or qq), except for $\mathrm{t\bar{t}H}$ which is considered separately in both cases.

Table 6.9 Cross section (rate) uncertainties assigned to MC simulated processes. Each column in the table is an independent source of uncertainty. Uncertainties in the same column for two different processes are completely correlated

Process	PDF				QCD Scale				
	$gg_{t\bar{t}H}$	gg	$q\bar{q}$	qg	$t\bar{t}$	t	V	VV	$t\bar{t}H$
$t\bar{t}H$	3.6%								+5.8% −9.2%
$t\bar{t}$ + jets		4.0%			+2.0% −4.0%				
$t\bar{t}$ + W			2.0%		+13% −12%				
$t\bar{t}$ + Z		3.0%			+10% −12%				
Single t				3.0%		+3.0% −2.0%			
W/Z + jets			4.0%				1.0%		
Diboson			2.0%					2.0%	

The variation in the MEM discriminant distributions due to the uncertainty in the PDF set was evaluated by using the different sub-PDFs of the employed `NNPDF3.0` PDF set, and was found to be negligible.

$t\bar{t}$+ *heavy-flavour cross sections* The $t\bar{t}$ + heavy-flavour processes represent important sources of irreducible background, which have not yet been measured, nor subjected to higher-order calculations that constrain these contributions. In fact, the most recent direct measurement of the $t\bar{t} + b\bar{b}$ cross section has an accuracy of ≈35% [26]. However, this measurement was made using a dilepton selection and so cannot be reliably used for the fully hadronic $t\bar{t} + b\bar{b}$ decay. Therefore, a conservative 50% uncertainty on the production rate is assigned separately to the $t\bar{t} + b\bar{b}$, $t\bar{t}$ + 2b, $t\bar{t}$ + b, and $t\bar{t} + c\bar{c}$ processes. These uncertainties are treated as uncorrelated in the final fit, and are in addition to the cross section uncertainties listed in Table 6.9. Ignoring the $t\bar{t}$ + heavy-flavour cross section uncertainties improves the expected exclusion limit (see Sect. 7.2) by around 5%.

MC statistics The limited number of simulated background and signal events leads to statistical fluctuations in the nominal prediction. This is taken into account by assigning a nuisance parameter for each bin of each sample that can be changed by its uncertainty as described in Ref. [27, 28]. This results in 1200 independent nuisance parameters across the six analysis categories (20 processes × 60 bins).

QCD Multijet Estimation

Many uncertainties that would be related to a MC simulation of the multijet background have been avoided by estimating its contribution from data. Nevertheless, a few small systematic uncertainties must be considered. First, all the uncertainties described above that affect simulated backgrounds are propagated to the multijet background when subtracting the simulated backgrounds from data in the control region.[4] Second, the statistical uncertainties from the data and limited MC simulation in the control region are carried over to the QCD multijet estimate and form the equivalent MC statistical uncertainty for the multijet background as described above for other samples. Finally, a number of uncertainties that exclusively affect the multijet process are described below.

Loose b-tagged jet correction The corrections to loose b-tagged jets used in the multijet estimation method show small dependencies between variables, as described in Sect. 6.4.2. The perfect correction would be obtained by repeatedly applying the correction for each variable after the other until a minimum deviation is observed. Since this process is truncated at a single correction for each variable, a systematic uncertainty is applied to the method, which is derived from the next correction to η. This re-correction is used to derive the "up" variation of the uncertainty while the "down" variation is set equal to the nominal correction.

MEM first bin A small systematic uncertainty is attributed to the consistent over or underestimation observed in the first bin of the MEM discriminant in the 4b and 3b categories of the validation regions, respectively. A 2.5% uncertainty is applied on the first bin, correlated across 4b categories, and a 2.0% uncertainty is applied on the first bin, correlated across 3b categories. The normalisation of the other nine bins are adjusted proportionally such that the total multijet yield is unchanged under these systematic variations.

H_T reweighting As mentioned in Sect. 6.4.2, two systematic uncertainties are applied to account for mismodelling at low H_T. A reweighting based on the H_T distribution (considering only the first 6 leading jets in p_T) is derived in the 7j, 3b category and in an inclusive 4 or more b-tag region and applied separately to the 7j, 3b and 4b categories, resulting in two uncorrelated uncertainties, the latter being 100% correlated across the three categories to which it applies. Each reweighting represents the "up" variation of the uncertainty, while the "down" variation is set equal to the nominal, i.e. without any H_T reweighting.

QCD multijet normalisation The total normalisation in each category is left unconstrained in the final fit. The uncertainties in multijet normalisation have the largest impact on the sensitivity of the analysis, and setting the normalisation to a fixed value in each category improves the expected limit by 20 to 30%.

In total, there are 58 independent sources of systematic uncertainty, plus 1200 separate bin-by-bin nuisance parameters, and six unconstrained normalisation parameters. A summary of the various sources and their impact on yields is provided in

[4] Rate uncertainties can affect the distribution of the multijet MEM discriminant through the subtraction from data in the control region. Therefore all rate uncertainties are implemented as shape uncertainties in the final fit, where the multijet rate and distribution change, but only the rates of affected simulated processes change.

Table 6.10 Summary of the systematic uncertainties affecting the signal and background expectations. The second column indicates the range in yield of affected processes caused by changing the nuisance parameters by their uncertainties. The third column indicates if the uncertainties change the distribution in the MEM discriminant. A checkmark (✓) indicates that the uncertainty applies to the stated processes. An asterisk (*) indicates that the uncertainty affects the data-based multijet estimate distribution indirectly through the subtraction of directly affected backgrounds in the control region

Source (number if > 1)	Range of uncertainty	Shape	Process			
			$t\bar{t}H$	Multijet	$t\bar{t}+$jets	Others
Experimental uncertainties						
Integrated luminosity	2.5%	No	✓	*	✓	✓
Trigger efficiency	1–2%	Yes	✓	*	✓	✓
Pileup	0.2–5%	Yes	✓	*	✓	✓
JES (25)	3–11%	Yes	✓	*	✓	✓
JER	2–11%	Yes	✓	*	✓	✓
b tagging (9)	4–40%	Yes	✓	*	✓	✓
QGL reweighting	4–11%	Yes	✓	*	✓	✓
Top quark p_T reweighting	1–2%	Yes	–	*	✓	–
QCD multijet estimation						
CSVL correction	–	Yes	–	✓	–	–
MEM first bin (2)	–	Yes	–	✓	–	–
H_T reweighting (2)	–	Yes	–	✓	–	–
Normalisation (6)	∞	No	–	✓	–	–
Theoretical uncertainties						
$t\bar{t}+b\bar{b}$ normalization	50%	No	–	*	✓	–
$t\bar{t}+2b$ normalization	50%	No	–	*	✓	–
$t\bar{t}+b$ normalization	50%	No	–	*	✓	–
$t\bar{t}+c\bar{c}$ normalization	50%	No	–	*	✓	–
QCD scale–signal	6–9%	No	✓	–	–	–
QCD scale–background (4)	1–13%	No	–	*	✓	✓
PDF (4)	2–4%	No	✓	*	✓	✓
MC statistics (1200)	2–40%	Yes	✓	*	✓	✓

Table 6.10. To give an indication of the effect of some of the systematic uncertainties, Fig. 6.19 shows the MEM distribution for the $t\bar{t}H$ ($H \to b\bar{b}$), $t\bar{t}+b\bar{b}$ and QCD multijet processes in the 8j, ≥4b category under systematic variations of the JES (all components considered together), JER, charm-flavour b-tagging scale factor, QGL reweighting and pileup reweighting uncertainties.

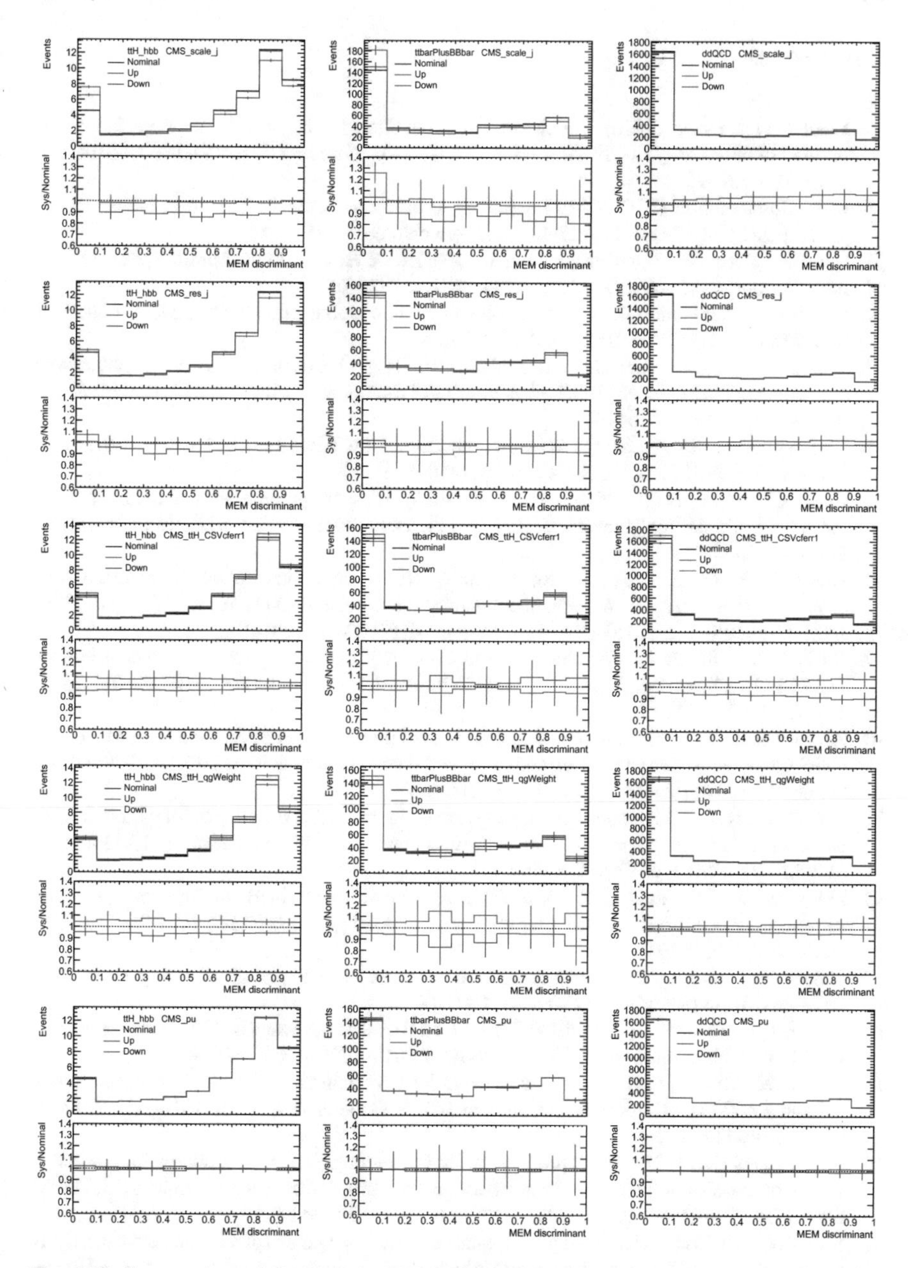

Fig. 6.19 Systematic variations of the MEM discriminant in the 8j, ≥4b category under JES (row 1), JER (row 2), charm-flavour b-tag scale factor (row 3), QGL reweighting (row 4) and pileup reweighting (row 5), for the $t\bar{t}H$ ($H \rightarrow b\bar{b}$) (left), $t\bar{t} + b\bar{b}$ (centre) and QCD multijet (right) processes. The nominal distribution is in black, the systematic variation "Up" is in red and the variation "Down" is in blue

References

1. CMS Collaboration (2018) Search for $\mathrm{t\bar{t}H}$ production in the all-jet final state in proton-proton collisions at $\sqrt{s} = 13$ TeV. JHEP 06:101. https://doi.org/10.1007/JHEP06(2018)101, arXiv:1803.06986
2. CMS Collaboration (2017) CMS luminosity measurements for the 2016 data taking period. CMS-PAS-LUM-17-001. https://www.cds.cern.ch/record/2257069
3. Sjöstrand T et al (2015) An introduction to PYTHIA 8.2. Comput Phys Commun 191:159–177. https://doi.org/10.1016/j.cpc.2015.01.024. arXiv:1410.3012
4. GEANT4 Collaboration (2003) GEANT4-a simulation toolkit. Nucl Instrum Meth A 506:250. https://doi.org/10.1016/S0168-9002(03)01368-8
5. Hartanto HB, Jager B, Reina L, Wackeroth D (2015) Higgs boson production in association with top quarks in the POWHEG BOX. Phys Rev D 91:094003. https://doi.org/10.1103/PhysRevD.91.094003, arXiv:1501.04498
6. Collaboration NNPDF (2015) Parton distributions for the LHC Run II. JHEP 04:040. https://doi.org/10.1007/JHEP04(2015)040. arXiv:1410.8849
7. Campbell JM, Ellis RK, Nason P, Re E (2015) Top-pair production and decay at NLO matched with parton showers. JHEP 04:114. https://doi.org/10.1007/JHEP04(2015)114. arXiv:1412.1828
8. Alioli S, Nason P, Oleari C, Re E (2009) NLO single-top production matched with shower in POWHEG: s- and t-channel contributions. JHEP 09:111. https://doi.org/10.1007/JHEP02(2010)011, arXiv:0907.4076. [Erratum: JHEP 02:011 (2010)]
9. Re E (2011) Single-top Wt-channel production matched with parton showers using the POWHEG method. Eur Phys J C 71:1547. https://doi.org/10.1140/epjc/s10052-011-1547-z. arXiv:1009.2450
10. Alwall J et al (2014) The automated computation of tree-level and next-to-leading order differential cross sections, and their matching to parton shower simulations. JHEP 07:079. https://doi.org/10.1007/JHEP07(2014)079. arXiv:1405.0301
11. Alwall J et al (2008) Comparative study of various algorithms for the merging of parton showers and matrix elements in hadronic collisions. Eur Phys J C 53:473. https://doi.org/10.1140/epjc/s10052-007-0490-5. arXiv:0706.2569
12. Mangano ML et al (2003) ALPGEN, a generator for hard multiparton processes in hadronic collisions. JHEP 07:001. https://doi.org/10.1088/1126-6708/2003/07/001. arXiv:hep-ph/0206293
13. CMS Collaboration (2014) Underlying event tunes and double parton scattering. CMS-PAS-GEN-14-001. https://cds.cern.ch/record/1697700
14. Skands P, Carrazza S, Rojo J (2014) Tuning PYTHIA 8.1: the Monash, (2013) tune. Eur Phys J C 74:3024. https://doi.org/10.1140/epjc/s10052-014-3024-y, arXiv:1404.5630
15. Czakon M, Mitov A (2014) Top++: a program for the calculation of the top-pair cross-section at hadron colliders. Comput Phys Commun 185:2930. https://doi.org/10.1016/j.cpc.2014.06.021. arXiv:1112.5675
16. Kidonakis N (2014) Top quark production. In: Proceedings, Helmholtz international summer school on physics of heavy quarks and hadrons (HQ 2013): JINR, Russia, July 2013. https://doi.org/10.3204/DESY-PROC-2013-03/Kidonakis, arXiv:1311.0283
17. Kant P et al (2015) HatHor for single top-quark production: updated predictions and uncertainty estimates for single top-quark production in hadronic collisions. Comput Phys Commun 191:74. https://doi.org/10.1016/j.cpc.2015.02.001. arXiv:1406.4403
18. Campbell JM, Ellis RK, Williams C (2011) Vector boson pair production at the LHC. JHEP 07:018. https://doi.org/10.1007/JHEP07(2011)018. arXiv:1105.0020
19. Maltoni F, Pagani D, Tsinikos I (2016) Associated production of a top-quark pair with vector bosons at NLO in QCD: impact on $\mathrm{t\bar{t}H}$ searches at the LHC. JHEP 02:113. https://doi.org/10.1007/JHEP02(2016)113, arXiv:1507.05640

20. LHC Higgs Cross Section Working Group Collaboration (2016) Handbook of LHC higgs cross sections: 4. deciphering the nature of the higgs sector. FERMILAB-FN-1025-T, CERN-2017-002-M, 10.23731/CYRM-2017-002. https://doi.org/10.23731/CYRM-2017-002, arXiv:1610.07922
21. Utilities for accessing pileup information for data. https://twiki.cern.ch/twiki/bin/view/CMS/PileupJSONFileforData#Pileup_JSON_Files_For_Run_II, r28. CMS Internal
22. CMS Collaboration (2017) Measurement of differential cross sections for top quark pair production and associated jets using the lepton+jets final state in proton-proton collisions at 13 TeV. CMS-PAS-TOP-17-002. https://cds.cern.ch/record/2284596
23. Event reweighting using scale factors calculated with a tag and probe method. https://twiki.cern.ch/twiki/bin/view/CMS/BTagShapeCalibration, r17. CMS Internal
24. Quark-gluon likelihood at 13 TeV. https://twiki.cern.ch/twiki/bin/view/CMS/QuarkGluonLikelihood#Systematics, r27. CMS Internal
25. Collaboration CMS (2011) Determination of jet energy calibration and transverse momentum resolution in CMS. JINST 6:P11002. https://doi.org/10.1088/1748-0221/6/11/P11002. arXiv:1107.4277
26. Collaboration CMS (2018) Measurements of $t\bar{t}$ cross sections in association with b jets and inclusive jets and their ratio using dilepton final states in pp collisions at $\sqrt{s} = 13$ TeV. Phys Lett B 776:355. https://doi.org/10.1016/j.physletb.2017.11.043, arXiv:1705.10141
27. Barlow RJ, Beeston C (1993) Fitting using finite Monte Carlo samples. Comput Phys Commun 77:219–228. https://doi.org/10.1016/0010-4655(93)90005-W
28. Conway JS (2011) Incorporating nuisance parameters in likelihoods for multisource spectra. In: Proceedings, PHYSTAT 2011 workshop on statistical issues related to discovery claims in search experiments and unfolding, CERN, January 2011, p 115, arXiv:1103.0354, https://doi.org/10.5170/CERN-2011-006.115

Chapter 7
Results and Combination

The search is performed as a binned shape analysis, in which the distribution of the MEM discriminant forms the basis of the statistical treatment. The results are interpreted in terms of the signal-strength modifier, which is defined as the ratio of the measured $t\bar{t}H$ production cross section σ to the standard model (SM) prediction for $m_H = 125\,\text{GeV}$:

$$\mu = \sigma/\sigma_{SM} \tag{7.1}$$

The statistical method used to calculate the results is the same as used in other CMS Higgs boson analyses. It is extensively documented in Ref. [1], and its main features, in the context of this analysis, are described below, while the results of applying it to the analysis are presented in Sect. 7.2. A demonstration of the statistical calculation is provided in Appendix B.

7.1 Statistical Tools

In a binned shape analysis, the expected number of signal events in one or multiple bins of the signal-extraction histogram is denoted s, while the expected number of background events is denoted b. With the definition of the signal strength (7.1), the *background-only* hypothesis corresponds to $\mu = 0$ and b events, while the SM *signal+background* hypothesis corresponds to $\mu = 1$ and $s + b$ events. In setting exclusion limits on the signal strength, the modified frequentist method, also known as CL_s, is used [2, 3].

7.1.1 The Likelihood

Predictions for the signal and background event yields depend on various systematic uncertainties that are accounted for by introducing a set of nuisance parameters θ,

D. Salerno, *The Higgs Boson Produced With Top Quarks in Fully Hadronic Signatures*, Springer Theses, https://doi.org/10.1007/978-3-030-31257-2_7

so that the expected number of signal and background events become functions of the nuisance parameters: $s(\theta)$ and $b(\theta)$. The expected number of events in a given bin $i = 1, 2, \ldots, N$ is expressed as:

$$\nu_i = \mu \cdot s_i(\theta) + b_i(\theta), \tag{7.2}$$

while the number of observed data events in the bin is n_i. A likelihood function $\mathcal{L}(\text{data}|\mu, \theta)$, which represents the likelihood of observing the data given the signal strength μ and the true values of the nuisance parameters θ, is then constructed:

$$\begin{aligned}\mathcal{L}(\text{data}|\mu, \theta) &= \text{Poisson}\big[\text{data}|\mu \cdot s(\theta) + b(\theta)\big] \cdot p(\tilde{\theta}|\theta) \\ &= \prod_{i=1}^{N} \frac{(\mu s + b)^{n_i}}{n_i!} e^{-(\mu s+b)} \cdot p(\tilde{\theta}|\theta),\end{aligned} \tag{7.3}$$

where $p(\tilde{\theta}|\theta)$ is the joint probability density function (pdf) for the nuisance parameters, and $\tilde{\theta}$ represents their default values. The signal strength is a free parameter in this model.

7.1.2 *Treatment of Systematic Uncertainties*

The systematic uncertainty pdfs in Eq. (7.3) are derived from the posterior pdfs $\rho(\theta|\tilde{\theta})$ through Bayes' theorem using flat hyper-priors. There are three different types of pdf relevant for this analysis:

- A uniform pdf is used for nuisance parameters that are unconstrained by any considerations or measurement not involving the data: $\rho(\theta|\tilde{\theta}) = c$.
- A Gaussian pdf is used for uncertainties on parameters that can take positive and negative values, or that are small relative to the value of the parameter:

$$\rho(\theta|\tilde{\theta}) = \frac{1}{\sqrt{2\pi}\sigma} \exp\left(-\frac{(\theta - \tilde{\theta})^2}{2\sigma^2}\right). \tag{7.4}$$

  Two perfectly correlated observables A and B with best estimates $\tilde{A}$ and $\tilde{B}$ can be generated from the same random variable X distributed as a standard normal, with $A = \tilde{A} \cdot (1 + \sigma_A \cdot X)$ and $B = \tilde{B} \cdot (1 + \sigma_B \cdot X)$. Perfect anti-correlations are considered by taking $\sigma_A > 0$ and $\sigma_B < 0$. This treatment of perfect correlations is used for different processes which are affected to different degrees by the same systematic uncertainty.
- A log-normal pdf is used for positively defined observables, such as cross sections, efficiencies or luminosity:

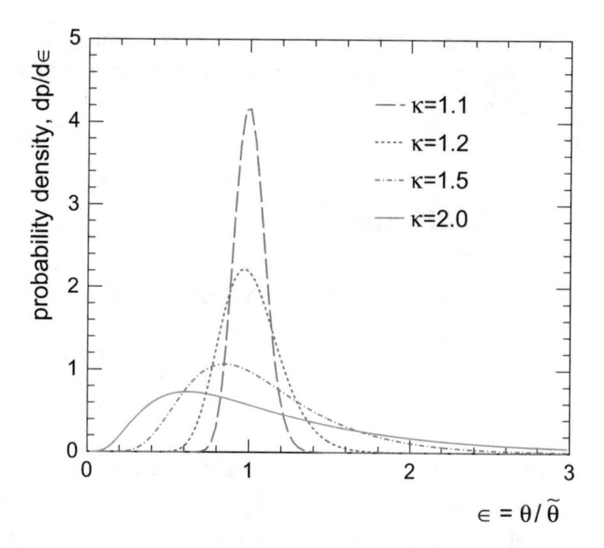

Fig. 7.1 Log-normal distribution with $\kappa = 1.1$, 1.2, 1.5 and 2.0

$$\rho(\theta|\tilde{\theta}) = \frac{1}{\sqrt{2\pi}\ln(\kappa)} \exp\left(-\frac{(\ln(\theta/\tilde{\theta}))^2}{2(\ln\kappa)^2}\right)\frac{1}{\theta}, \tag{7.5}$$

where the width of the distribution is characterised by $\kappa = 1 + \epsilon$, with ϵ the relative uncertainty. For small uncertainties, the Gaussian and log-normal are asymptotically identical, while the log-normal is more appropriate for large uncertainties, e.g. it nicely accommodates a factor-of-2 uncertainty. Figure 7.1 displays log-normal distributions for a few different values of κ. Compared to a Gaussian, the log-normal has a longer tail and goes to zero at $\theta = 0$. In analogy to the above case, two perfectly correlated observables A and B with best estimates $\tilde{A}$ and $\tilde{B}$, and log-normal uncertainties κ_A and κ_B, can be generated from the same random variable X distributed as a standard normal, by taking $A = \tilde{A} \cdot \kappa_A^X$ and $B = \tilde{B} \cdot \kappa_B^X$. Perfect anti-correlations are considered by taking $\kappa_A > 0$ and $\kappa_B < 0$.

There are three types of systematic uncertainty considered in the analysis, each of which uses one or more of the pdfs above:

- Rate parameters are used for the QCD multijet normalisation. No prior information on the magnitude of the normalisation is assumed, nor is any constraint placed on it. The rate parameters use the uniform pdf.
- Rate uncertainties are expressed in terms of a relative uncertainty and use the log-normal pdf.
- Shape uncertainties are modelled by defining two additional histograms, corresponding to one-standard-deviation shifts up and down of the uncertainty: $\zeta = -1, 0, 1$, where $\zeta = 0$ is the nominal histogram. A family of histograms is then derived by a bin-by-bin quadratic interpolation of the three base histograms within $|\zeta| < 1$, and linear extrapolation for $|\zeta| > 1$. The pure shape component is

considered by first normalising the histograms, and then using a Gaussian pdf for the parameter ζ. Any rate component of the uncertainty is treated separately under a log-normal pdf.

7.1.3 Limit Setting

To calculate the upper exclusion limits, the profile likelihood ratio is used:

$$\lambda(\mu) = \frac{\mathcal{L}(\text{data}|\mu, \hat{\theta}_\mu)}{\mathcal{L}(\text{data}|\hat{\mu}, \hat{\theta})} \tag{7.6}$$

where $\hat{\theta}_\mu$ represents the conditional maximum likelihood estimator of θ, given a fixed signal strength μ and the data: $\hat{\theta}_\mu = \theta(\mu, \text{data})$. On the other hand, $\hat{\mu}$ and $\hat{\theta}$ represent the values of the signal strength and nuisance parameters at the global maximum of the likelihood, Eq. (7.3). A test statistic is then used to compare competing hypotheses for the value of μ:

$$\tilde{q}_\mu = -2 \ln \frac{\mathcal{L}(\text{data}|\mu, \hat{\theta}_\mu)}{\mathcal{L}(\text{data}|\hat{\mu}, \hat{\theta})}, \qquad 0 \leq \hat{\mu} \leq \mu, \tag{7.7}$$

The lower constraint of $0 \leq \hat{\mu}$ is required by physics, i.e. the signal rate cannot be negative, while the upper constraint of $\hat{\mu} \leq \mu$ is imposed to guarantee a one-sided confidence interval, i.e. not detached from zero. The physical implication is that upward fluctuations of the data, such that $\hat{\mu} > \mu$, are not considered as evidence against the signal+background hypothesis, i.e. a signal with strength μ.

From Eq. (7.6) it is clear that $0 < \lambda \leq 1$, and values of λ close to 1 imply a good agreement between data and the value of μ being tested. The value of the test statistic (7.7) is therefore positive, with higher values implying poorer agreement, as shown in Fig. 7.2.

The next stage of the procedure involves constructing the pdf of the test statistic under the signal+background and background-only hypothesis:

$$f(\tilde{q}_\mu|\mu, \hat{\theta}_\mu), \qquad f(\tilde{q}_\mu|0, \hat{\theta}_0). \tag{7.8}$$

This can be accomplished by generating many toy Monte Carlo (MC) pseudo data for $\tilde{q}_\mu$ under the two different hypotheses, where the values of the nuisance parameters are initially set to their conditional maximum likelihood estimates, $\hat{\theta}_\mu$ and $\hat{\theta}_0$, using the real data. An alternative method, valid for large datasets and widely used at the LHC, is based on the asymptotic approximation [4]:

$$-2 \ln \lambda(\mu) \approx \frac{(\mu - \hat{\mu})^2}{\sigma^2}, \tag{7.9}$$

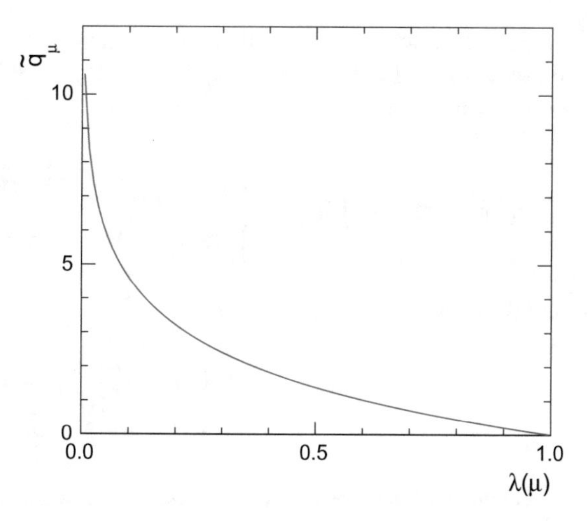

Fig. 7.2 Distribution of the test statistic $\tilde{q}_\mu = -2 \ln \lambda(\mu)$

where $\hat{\mu}$ follows a Gaussian distribution with mean μ', equal to the signal strength of the data, and standard deviation σ. The latter can be estimated using the Asimov data set,[1] i.e. the pseudo data set equal to the expected background with the nominal nuisance parameters, thus giving maximum likelihood estimators of $\hat{\mu} = 0$ and $\hat{\theta} = \tilde{\theta}$. If the test statistic of the Asimov data set is $q_{\mu,A}$, then:

$$\sigma^2 = \frac{\mu^2}{q_{\mu,A}}. \tag{7.10}$$

Under these assumptions, with $\mu = \mu'$, the distribution of the test statistic follows an analytical expression:

$$f(\tilde{q}_\mu|\mu) = \frac{1}{2}\delta(\tilde{q}_\mu) + \begin{cases} \frac{1}{2}\frac{1}{\sqrt{2\pi}}\frac{1}{\sqrt{\tilde{q}_\mu}}e^{-\tilde{q}_\mu/2} & 0 < \tilde{q}_\mu \leq \mu^2/\sigma^2 \\ \frac{1}{\sqrt{2\pi}(2\mu/\sigma)}\exp\left[-\frac{1}{2}\frac{(\tilde{q}_\mu+\mu^2/\sigma^2)^2}{(2\mu/\sigma)^2}\right] & \tilde{q}_\mu > \mu^2/\sigma^2, \end{cases} \tag{7.11}$$

where the delta function ensures a probability of 0.5 at $\tilde{q}_\mu = 0$. The corresponding cumulative distribution function (CDF) is given by:

$$F(\tilde{q}_\mu|\mu) = \begin{cases} \Phi\left(\sqrt{\tilde{q}_\mu}\right) & 0 < \tilde{q}_\mu \leq \mu^2/\sigma^2 \\ \Phi\left(\frac{(\tilde{q}_\mu+\mu^2/\sigma^2)}{(2\mu/\sigma)}\right) & \tilde{q}_\mu > \mu^2/\sigma^2, \end{cases} \tag{7.12}$$

where Φ is the CDF of the standard normal distribution.

[1]The Asimov dataset can in general be set equal to the signal+background expectation of arbitrary μ. In this case the maximum likelihood estimators are $\hat{\mu} = \mu$ and $\hat{\theta} = \tilde{\theta}$.

The asymptotic approximation can be made for the background-only hypothesis by setting $\mu' = 0$. In this case the pdf of the test statistic is given by a slightly more complex formula:

$$f(\tilde{q}_\mu|0) = \Phi\left(\frac{-\mu}{\sigma}\right)\delta(\tilde{q}_\mu) + \begin{cases} \frac{1}{2}\frac{1}{\sqrt{2\pi}}\frac{1}{\sqrt{\tilde{q}_\mu}}\exp\left[-\frac{1}{2}\left(\sqrt{\tilde{q}_\mu} - \frac{\mu}{\sigma}\right)^2\right] & 0 < \tilde{q}_\mu \leq \mu^2/\sigma^2 \\ \frac{1}{\sqrt{2\pi}(2\mu/\sigma)}\exp\left[-\frac{1}{2}\frac{(\tilde{q}_\mu - \mu^2/\sigma^2)^2}{(2\mu/\sigma)^2}\right] & \tilde{q}_\mu > \mu^2/\sigma^2, \end{cases} \tag{7.13}$$

and the CDF is given by:

$$F(\tilde{q}_\mu|0) = \begin{cases} \Phi\left(\sqrt{\tilde{q}_\mu} - \frac{\mu}{\sigma}\right) & 0 < \tilde{q}_\mu \leq \mu^2/\sigma^2 \\ \Phi\left(\frac{(\tilde{q}_\mu - \mu^2/\sigma^2)}{(2\mu/\sigma)}\right) & \tilde{q}_\mu > \mu^2/\sigma^2. \end{cases} \tag{7.14}$$

Given the data, the observed value of the test statistic $\tilde{q}_\mu^{\text{obs}}$ and the maximum likelihood estimates of the nuisance parameters $\hat{\theta}_\mu^{\text{obs}}$ can be calculated. With these, two p-values for the observation under the signal+background and background-only hypotheses are defined:

$$p_\mu = P(\tilde{q}_\mu \geq \tilde{q}_\mu^{\text{obs}}|\mu s + b) = \int_{\tilde{q}_\mu^{\text{obs}}}^{\infty} f(\tilde{q}_\mu|\mu, \hat{\theta}_\mu^{\text{obs}})\mathrm{d}\tilde{q}_\mu \quad = 1 - F(\tilde{q}_\mu^{\text{obs}}|\mu, \hat{\theta}_\mu^{\text{obs}}), \tag{7.15}$$

$$1 - p_b = 1 - P(\tilde{q}_\mu < \tilde{q}_\mu^{\text{obs}}|b) = \int_{\tilde{q}_\mu^{\text{obs}}}^{\infty} f(\tilde{q}_\mu|0, \hat{\theta}_0^{\text{obs}})\mathrm{d}\tilde{q}_\mu \quad = 1 - F(\tilde{q}_\mu^{\text{obs}}|0, \hat{\theta}_0^{\text{obs}}). \tag{7.16}$$

The ratio of these two probabilities is used to calculate the modified frequentist statistic:

$$\text{CL}_s(\mu) = \frac{p_\mu}{1 - p_b}, \tag{7.17}$$

which is in turn used to calculate the confidence level (CL) of exclusion. If $\text{CL}_s(\mu) \leq \alpha$, then the signal with strength μ is said to be excluded at the $(1 - \alpha)$ CL. For the often quoted 95% CL upper limit on μ, the value of μ is adjusted until $\text{CL}_s(\mu) = 0.05$. The modified frequentist statistic gives one-sided exclusion limits by construction and protects from under-fluctuations of the background in the presence of a zero or weak signal strength.

The expected upper limit, given the background-only hypothesis, is a useful metric for the sensitivity of a search. It is calculated by generating MC pseudo data with the background-only hypothesis or by using the asymptotic approximation above, Eq. (7.11), and the Asimov dataset. The median expected limit is often quoted with its 68% and 95% uncertainty bands, in which the observed limit is expected to lie under the background-only hypothesis.

7.1.4 *Significance of an Excess*

In the case of observed data above the background-only expectation, the significance of the excess is quantified by the p-value under the background-only hypothesis. This equates to the probability of a background fluctuation that gives an excess equal to or greater than that observed. In this case, the following test statistic is used:

$$q_0 = -2\ln\frac{\mathcal{L}(\text{data}|0,\hat{\hat{\theta}}_0)}{\mathcal{L}(\text{data}|\hat{\mu},\hat{\theta})}, \qquad \hat{\mu} \geq 0, \tag{7.18}$$

where the constraint $\hat{\mu} \geq 0$ prevents a deficit of events with the respect to the background-only hypothesis from resulting in a high significance. As for $\tilde{q}_\mu$, the pdf of q_0 is constructed with either toy pseudo data or the asymptotic approximation and the Asimov data set. In the latter case, the pdf is given by [4]:

$$f(q_0|0) = \frac{1}{2}\delta(\tilde{q}_0) + \frac{1}{2}\frac{1}{\sqrt{2\pi}}\frac{1}{\sqrt{q_0}}e^{-q_0/2}, \tag{7.19}$$

and the CDF by:

$$F(q_0|0) = \Phi\left(\sqrt{q_0}\right). \tag{7.20}$$

The p-value of the observed data q_0^{obs} is then calculated as:

$$p_0 = P(q_0 \geq q_0^{\text{obs}}|b) = \int_{q_0^{\text{obs}}}^{\infty} f(q_0|0,\hat{\hat{\theta}}_0^{\text{obs}})\text{d}q_0 = 1 - F(q_0^{\text{obs}}|0), \tag{7.21}$$

which is then converted into a significance in terms of the number of standard deviations of a standard normal distribution:

$$Z = \Phi^{-1}(1 - p_0). \tag{7.22}$$

A 5σ significance ($Z = 5$) is generally required to claim a discovery, which corresponds to a p-value of $p_0 = 2.87 \times 10^{-7}$.

7.2 Analysis Results

The tools outlined in Sect. 7.1 have been applied to the combination of all six analysis categories, considering all systematic uncertainties. For an indication of the contribution of each category to the total sensitivity, the procedure has also been applied separately to the individual categories and to smaller combinations of just the 3b and the 4b categories. The results of each type of statistical test are presented below.

7.2.1 Maximum Likelihood Fit

The likelihood in Eq. (7.3) is maximised by performing a fit to the data with a floating signal strength and the nuisance parameters floating according to their assigned pdfs. In total, there are 60 bins of the MEM discriminant (10 in each category), and 1264 nuisance parameters.

The maximum likelihood estimator of the signal strength is $\hat{\mu} = 0.9 \pm 1.5$, while the post-fit values of the Gaussian and log-normal nuisance parameters (except the MC statistical uncertainties) are shown in Fig. 7.3. The central value of each nuisance parameter θ_k is given in terms of the shift from its nominal value $\tilde{\theta}_k$ expressed in units of its nominal uncertainty (pull). The uncertainty of each is shown relative to its nominal uncertainty, which by definition is ± 1 on the same scale, and gives an indication of the constraint achieved by the data. The post-fit values of the QCD multijet normalisation and their uncertainties are shown in Table 7.1, expressed relative to the initial normalisation. The central values are not meaningful as other nuisance parameters can affect the multijet normalisation. On the other hand, the post-fit uncertainties give a useful indication of how well this background is constrained by the data.

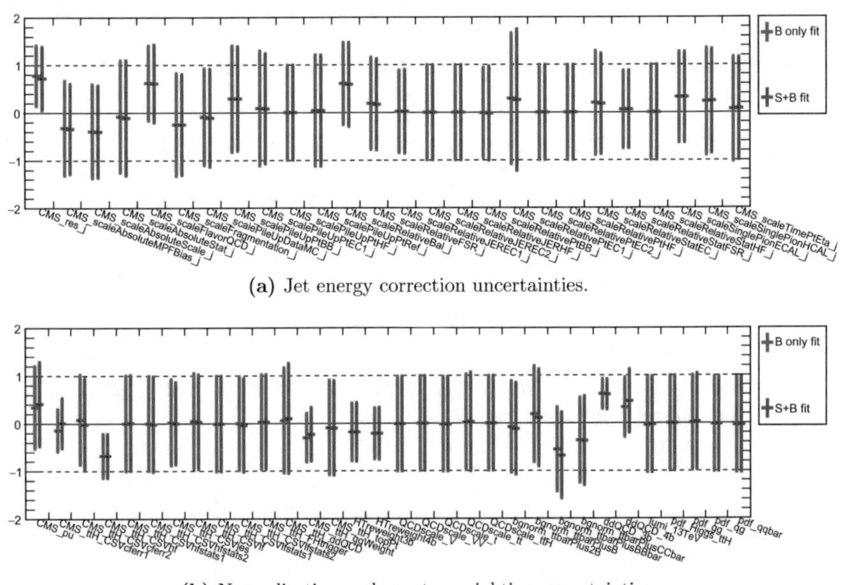

(a) Jet energy correction uncertainties.

(b) Normalisation and event reweighting uncertainties.

Fig. 7.3 Post-fit values of the nuisance parameters after the maximum likelihood fit to the data, for all systematic uncertainties except MC statistics and multijet normalisation. The background-only (B only) fit (blue) is made by ignoring the signal contribution, while the signal+background (S+B) fit (red) is made assuming a floating signal strength

Table 7.1 Post-fit values of the QCD multijet normalisation rate parameters. The maximum likelihood estimates are expressed relative to the initial normalisation in each category, which is by definition $1 \pm \infty$

Category	Normalisation
7j, 3b	1.005 ± 0.010
8j, 3b	1.005 ± 0.013
≥9j, 3b	1.008 ± 0.020
7j, ≥4b	1.012 ± 0.028
8j, ≥4b	1.010 ± 0.032
≥9j, ≥4b	1.031 ± 0.053

Discussion of Post-fit Parameters

The best fit value of the signal strength, $\hat{\mu} = 0.9 \pm 1.5$, is close to the SM prediction of $\mu = 1$. It is certainly compatible with the SM expectation given the relatively large uncertainties. This is the first indication that the predictions of the SM background and signal adequately describe the data.

The best fit values of the nuisance parameters are all contained within their pre-fit uncertainties. Most are constrained by the data, as indicated by the reduced size of the uncertainties in Fig. 7.3, while a few have post-fit uncertainties larger than 1, indicating that their value is not well determined by the data. A discussion of the largest pulls under the signal+background fit and their values follows:

- `CMS_res_j`: 0.72 ± 0.66. The jet energy resolution (JER) uncertainty is pulled 72% toward the "up" variation and constrained to 66% of the nominal uncertainty. Since this uncertainty is common to all CMS searches involving jets, a comparison with other analyses can indicate whether the nominal JER is a genuine underestimate of the true JER, or if the pull is a random fluctuation of this analysis. Indeed, the JER nuisance parameter is pulled both up and down in other Higgs boson searches. Despite the large pull of this parameter, varying it within its post-fit uncertainty only contributes about 0.25 to the total uncertainty on $\hat{\mu}$.
- `CMS_ttH_CSVhf`: -0.68 ± 0.46. The uncertainty on the b-tagging scale factors caused by the heavy-flavour jet contamination of the light-flavour jet control sample is pulled 68% toward the "down" variation, and is constrained to 46% of its nominal uncertainty. This uncertainty is unique to the two $\mathrm{t\bar{t}H}$ ($\mathrm{H \to b\bar{b}}$) searches at CMS, the other search observing an opposite pull with a similar constraint. The ultimate impact of this uncertainty on $\hat{\mu}$ is around 0.05.
- `bgnorm_ttbarPlusBBbar`: -0.68 ± 0.90. The additional 50% uncertainty on the $\mathrm{t\bar{t}} + \mathrm{b\bar{b}}$ cross section is pulled down by 68% indicating an over estimate of the cross section. Its uncertainty is slightly constrained to 90% of the nominal value, indicating that the initial uncertainty is justified. The other CMS $\mathrm{t\bar{t}H}$ ($\mathrm{H \to b\bar{b}}$) search observed a slight downward pull and a stronger constraint. The impact of this uncertainty is among the largest, contributing around 0.3 to the total uncertainty

on $\hat{\mu}$. This uncertainty is pulled down slightly less under the background-only fit, since the $\mathrm{t\bar{t}+b\bar{b}}$ contribution in signal-rich categories is high. It also represents a large loss of sensitivity for the other $\mathrm{t\bar{t}H}$ ($\mathrm{H \to b\bar{b}}$) analysis and is the subject of ongoing efforts to constrain the $\mathrm{t\bar{t}+b\bar{b}}$ cross section.

- `CMS_scaleFlavorQCD_j`: 0.61 ± 0.81. The uncertainty on the jet energy scale (JES) caused by the QCD flavour uncertainty in the simulated control samples is pulled up by 61% and constrained to 81% of its pre-fit value. This uncertainty is common to all CMS searches involving jets, however only a few searches use the factorised JES uncertainties, and thus have this nuisance parameter. The other CMS $\mathrm{t\bar{t}H}$ ($\mathrm{H \to b\bar{b}}$) search observed an opposite pull and a much stronger constraint. Despite the large pull of this uncertainty, its impact on $\hat{\mu}$ is only around 0.1.
- `ddQCD_3b`: 0.60 ± 0.31. The uncertainty on the first bin of the multijet MEM discriminant in the 3b categories is pulled up by 60%. It is highly constrained by the data, resulting in a post-fit uncertainty of just 31% of the pre-fit value. Nevertheless, its impact on $\hat{\mu}$ is less than 0.1.
- `CMS_scaleRelativeBal_j`: 0.59 ± 0.88. The uncertainty on the JES caused by the relative balance uncertainty in the simulated control samples is pulled up by 59% and constrained to 88% of its pre-fit value. The other CMS $\mathrm{t\bar{t}H}$ ($\mathrm{H \to b\bar{b}}$) search observed an opposite pull and a stronger constraint. The impact on $\hat{\mu}$ caused by varying this nuisance parameter within its post-fit uncertainty is less than 0.2.

The uncertainty with the largest impact on $\hat{\mu}$ is the charm-flavour jet component of the b-tagging scale factor, `CMS_ttH_CSVcferr1`, with a post-fit value of 0.01 ± 0.52. Since the nominal uncertainty applied for this is quite conservative, the large constraint of around 50% is somewhat expected. Its contribution to the total uncertainty on $\hat{\mu}$ of around 0.7, means that this scale factor should not be allowed to be determined in the final fit, but should be better estimated to begin with. This is one area of improvement for future versions of this analysis.

The QCD multijet normalisation uncertainties are quite well constrained by the data, as shown in Table 7.1. The largest post-fit uncertainty is 5% in the ≥9j, ≥4b category, and the smallest is 1% in the high-statistics 7j, 3b category. In general, the uncertainty is smaller for the categories with higher event yields (cf. Table 6.8), although it is probably better described as increasing with increasing S/B. The impacts of these uncertainties are not easily determined, since the multijet normalisation is affected by most other systematic uncertainties, and thus the rate parameters are strongly correlated or anti-correlated with many other nuisance parameters. An estimate of the combined impact of all six rate parameters is obtained by comparing the uncertainty on $\hat{\mu}$ without any nuisance parameters and with only the rate parameters. This results in a contribution to the total uncertainty on $\hat{\mu}$ of around 0.5 for all multijet normalisation uncertainties combined. This is another area of improvement for the future of this analysis.

Other systematic uncertainties with large impacts on the signal strength are the multijet first bin uncertainty in the 4b categories, `ddQCD_4b`, and the uncertainty on the quark-gluon likelihood (QGL) reweighting, `CMS_ttH_qgWeight`. The former has a post-fit value of 0.46 ± 0.66 with an impact on $\hat{\mu}$ of around 0.45, and relates to the imperfect modelling of the QCD multijet background. The latter relates to the modelling of the QGL and affects all CMS searches employing the QGL in some way. It has a post-fit value of -0.23 ± 0.56 and an impact on $\hat{\mu}$ of around 0.3. Both of these uncertainties represent areas of improvement for this analysis, especially since few other analyses currently use the QGL.

7.2.2 *Post-fit Distributions*

After determining the maximum likelihood estimates of the signal strength and nuisance parameters, the estimated contribution of each background and signal process can be derived. These best fit contributions can be seen in the post-fit distributions of the MEM discriminant in each category, following the combined fit to all categories, which are shown in Fig. 7.4. As seen in the figures, the agreement between prediction and data is excellent, with only a single bin outside of the post-fit uncertainties. This 10th bin of the 7j, ≥4b category pulls the signal strength up in the maximum likelihood fit, as discussed in Sect. 7.2.3. A comparison to Figure 6.17 demonstrates the degree to which the total uncertainties are constrained by the data, and highlights the better agreement between prediction and data following the fit.

The post-fit event yields expected for the signal and the different background processes for an integrated luminosity of $35.9\,\text{fb}^{-1}$, and the yields observed in data, are listed in Table 7.2 for each analysis category. The post-fit yields are also displayed in Figured 7.5. In comparison to the pre-fit event yields in Table 6.8, the effect of the fit is to increase the total background contribution slightly to compensate for the reduced signal. In addition, the background composition is altered by the fit, giving more QCD multijet and $\mathrm{t\bar{t}}+\mathrm{lf}$, and less $\mathrm{t\bar{t}}+\mathrm{b\bar{b}}$ across all categories, while the change in other $\mathrm{t\bar{t}}+\mathrm{hf}$ processes varies from category to category. The yields from the minor background processes are also altered slightly.

For a better visualisation of the signal, an accumulation of the signal-rich bins of the MEM discriminant can be made. If S and B represent the signal and background yields in each bin of the MEM discriminant, then $r = S/B$ is a measure of the signal contribution in each bin. For the 60 bins across the six categories, r ranges from 0.001 to 0.053. The logarithm $\log_{10}(r)$ has a smoother distribution across its range of -2.9 to -1.3, and is thus more useful for visualisation. Simply plotting a histogram of the 60 values of $\log_{10}(r)$ gives an indication of the distribution of MEM discriminant bins, but not of actual events. Instead, a histogram in the $\log_{10}(r)$ variable is made where each entry is weighted by the yield of the process or data in that bin. In this way, the variable $\log_{10}(S/B)$ is constructed, from which distributions can be made for each signal and background process and the data. For an illustration of the SM signal contribution, the fit to the MEM discriminant is performed with a constraint in the signal strength of $\mu = 1$. Given that the maximum likelihood estimate of $\hat{\mu} = 0.9$ is close to the SM expectation, the effect on the background contributions should be minimal and thus the change in μ will dominate the impact on the $\log_{10}(S/B)$ variable. Figure 7.6 shows the distribution of $\log_{10}(S/B)$, where S/B is the ratio of

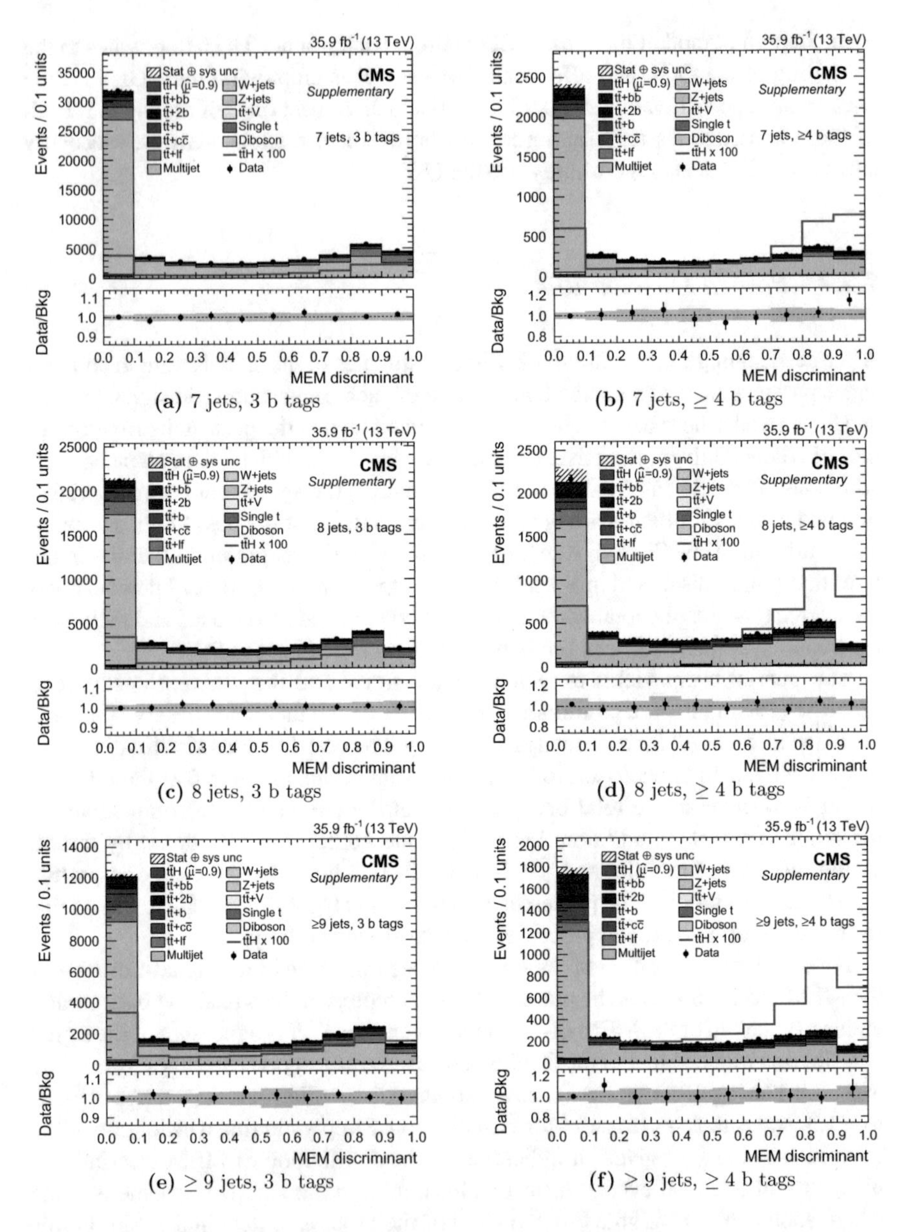

(a) 7 jets, 3 b tags

(b) 7 jets, ≥ 4 b tags

(c) 8 jets, 3 b tags

(d) 8 jets, ≥ 4 b tags

(e) ≥ 9 jets, 3 b tags

(f) ≥ 9 jets, ≥ 4 b tags

Fig. 7.4 Distributions in MEM discriminant for each analysis category after the combined fit to data. The fitted contributions expected from signal and background processes (filled histograms) are shown stacked. The signal distributions (lines) for a Higgs boson mass of $m_{\mathrm{H}} = 125\,\mathrm{GeV}$ are multiplied by a factor of 100 and superimposed on the data. Each background contribution is initially normalised to an integrated luminosity of $35.9\,\mathrm{fb}^{-1}$, while the multijet contribution of each category is free to float in the fit. The distributions observed in data (data points) are also shown. The ratios of data to background are given below the main panel

Table 7.2 Expected number of $t\bar{t}H$ signal and background events, and the observed event yields for the six analysis categories, after the fit to data. The quoted uncertainties contain all post-fit uncertainties described in Sect. 6.6 added in quadrature, considering all correlations among processes. The signal (S) and total background (B) ratios for the SM $t\bar{t}H$ expectation ($\mu = 1$) are also shown

Process	7j, 3b	8j, 3b	≥9j, 3b	7j, ≥4b	8j, ≥4b	≥9j, ≥4b
Multijet	47572 ± 2951	32713 ± 2221	17583 ± 1594	3531 ± 271	3768 ± 360	2279 ± 294
$t\bar{t}$ + lf	7678 ± 1568	5744 ± 1064	3164 ± 554	312 ± 127	408 ± 221	244 ± 96
$t\bar{t}$ + $c\bar{c}$	3055 ± 1404	2822 ± 1236	2170 ± 967	185 ± 103	272 ± 153	272 ± 153
$t\bar{t}$ + b	1395 ± 623	1235 ± 616	893 ± 424	142 ± 80	163 ± 109	134 ± 73
$t\bar{t}$ + 2b	894 ± 454	761 ± 370	599 ± 290	87 ± 58	114 ± 77	101 ± 52
$t\bar{t}$ + $b\bar{b}$	870 ± 340	1009 ± 367	969 ± 376	203 ± 90	366 ± 150	410 ± 168
Single t	745 ± 190	517 ± 129	284 ± 75	43 ± 20	78 ± 68	35 ± 17
W+jets	385 ± 167	210 ± 111	175 ± 216	30 ± 33	34 ± 110	4 ± 10
Z+jets	75 ± 21	82 ± 24	61 ± 19	6 ± 4	10 ± 6	13 ± 6
$t\bar{t}$ +V	110 ± 20	122 ± 27	120 ± 30	14 ± 7	28 ± 14	28 ± 14
Diboson	14 ± 5	5 ± 4	1 ± 1	0.6 ± 0.5	0.6 ± 0.6	0 ± 10
Total bkg	62792 ± 899	45221 ± 846	26020 ± 635	4553 ± 175	5241 ± 342	3522 ± 185
$t\bar{t}H$ ($\hat{\mu} = 0.9$)	130 ± 207	136 ± 218	118 ± 187	32 ± 51	46 ± 75	48 ± 77
Data	62920	45359	26136	4588	5287	3566
S/B ($\mu = 1$)	0.002	0.003	0.005	0.008	0.010	0.014
$S/\sqrt{B}$	0.59	0.70	0.79	0.52	0.69	0.85

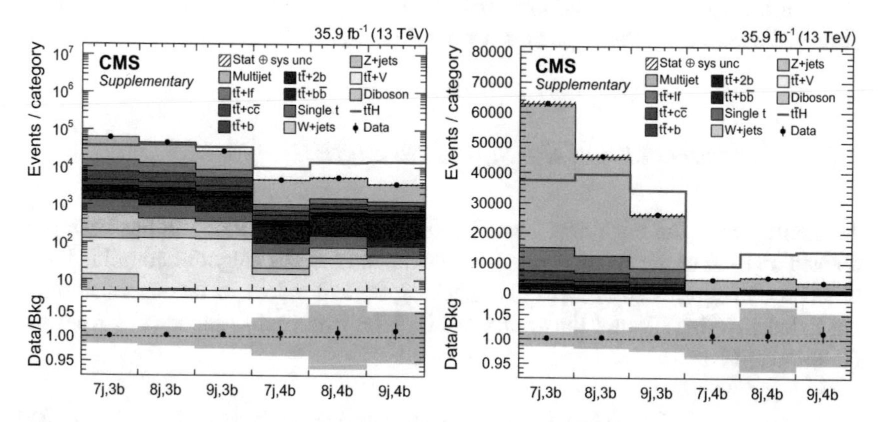

Fig. 7.5 Predicted (histograms) and observed (data points) event yields in each analysis category after the fit to data, corresponding to an integrated luminosity of 35.9 fb^{-1}. The expected contributions from different background processes (filled histograms) are stacked, showing the total fitted uncertainty (striped error bands), and the expected signal distribution (line) for a Higgs boson mass of $m_H = 125$ GeV is scaled to the total background yield for ease of readability. The ratios of data to background are given below the main panels, with the full uncertainties. The yields are shown with a logarithmic scale (left) and linear scale (right)

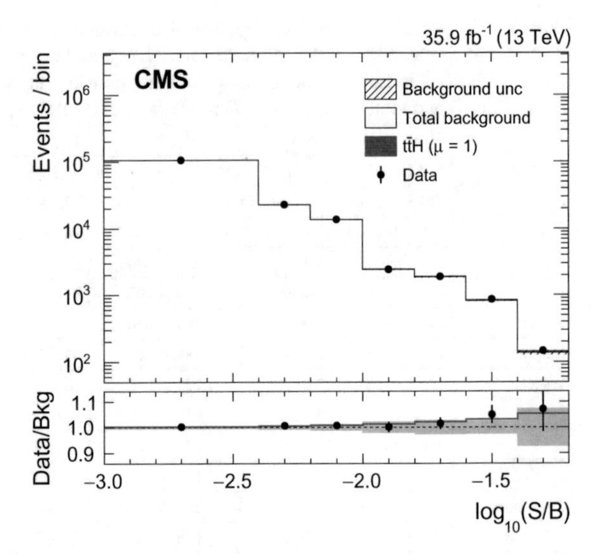

Fig. 7.6 Distribution in the logarithm $\log_{10}(S/B)$, where S and B indicate the respective bin-by-bin yields of the signal and background expected in the MEM discriminant distributions, as obtained from a combined fit with the constraint in the cross section of $\mu = 1$

the signal and background yields in each bin of the six MEM discriminant histograms, as obtained from a combined fit with the constraint in the signal strength of $\mu = 1$. Good agreement between the data and the SM signal+background expectation is observed over the whole range of this variable.

7.2.3 *Measurement of the Signal Strength*

As mentioned at the beginning Sect. 7.2, the maximum likelihood fit has been performed on each of the six categories, on the 3b and the 4b categories combined, and on all six categories combined. The resulting best-fit values of the signal strength are listed in Table 7.3 and illustrated in Fig. 7.7a. The best-fit value of μ from the combined fit is:

$$\hat{\mu} = 0.9 \pm 1.5 = 0.9 \pm 0.7\,(\text{stat}) \pm 1.3\,(\text{syst}), \tag{7.23}$$

where the total uncertainty is broken down into its statistical and systematic components. The statistical component is estimated by including in the fit only those nuisance parameters that are statistical in nature, namely the per-category multijet normalisations and the multijet bin-by-bin uncertainties, which are dominated by the uncertainty in the data in the control region. The systematic component is then calculated as the difference in quadrature between the total uncertainty and the statistical component. Since the multijet normalisations are correlated with many other systematic uncertainties, part of them is attributed to the systematic component and only part is reflected in the statistical component of the total uncertainty on $\hat{\mu}$.

Table 7.3 Best fit value of the signal-strength modifier μ and the median expected and observed 95% CL upper limits (UL) in each of the six analysis categories, as well as the combined results. The best fit values are shown with their total uncertainties and the breakdown into the statistical and systematic components. The expected limits are shown together with their 68% CL intervals

Category	Best fit μ and uncertainty			Observed	Expected
	$\hat{\mu}$	tot	(stat syst)	UL	UL
7j, 3b	1.6	$^{+9.6}_{-12.0}$	$\left(^{+2.7}_{-2.7}\ ^{+9.2}_{-11.7}\right)$	18.7	$17.6^{+6.2}_{-4.4}$
8j, 3b	1.2	$^{+5.9}_{-6.4}$	$\left(^{+2.2}_{-2.3}\ ^{+5.4}_{-5.9}\right)$	12.3	$11.5^{+4.6}_{-3.1}$
≥9j, 3b	−3.5	$^{+5.9}_{-6.5}$	$\left(^{+2.4}_{-2.4}\ ^{+5.4}_{-6.0}\right)$	9.0	$10.7^{+4.5}_{-3.1}$
7j, ≥4b	5.4	$^{+2.9}_{-2.7}$	$\left(^{+1.8}_{-1.8}\ ^{+2.3}_{-2.1}\right)$	10.6	$5.7^{+2.6}_{-1.7}$
8j, ≥4b	−0.2	$^{+2.8}_{-3.0}$	$\left(^{+1.5}_{-1.5}\ ^{+2.3}_{-2.6}\right)$	5.5	$5.5^{+2.6}_{-1.6}$
≥9j, ≥4b	−0.4	$^{+2.1}_{-2.2}$	$\left(^{+1.4}_{-1.3}\ ^{+1.6}_{-1.8}\right)$	4.0	$4.3^{+1.9}_{-1.3}$
3b categories	−1.7	$^{+5.2}_{-5.4}$	$\left(^{+1.4}_{-1.4}\ ^{+5.0}_{-5.3}\right)$	8.7	$9.2^{+3.7}_{-2.5}$
4b categories	1.5	$^{+1.6}_{-1.6}$	$\left(^{+0.9}_{-0.9}\ ^{+1.4}_{-1.4}\right)$	4.5	$3.3^{+1.5}_{-1.0}$
Combined	0.9	$^{+1.5}_{-1.5}$	$\left(^{+0.7}_{-0.7}\ ^{+1.3}_{-1.3}\right)$	3.8	$3.1^{+1.4}_{-0.9}$

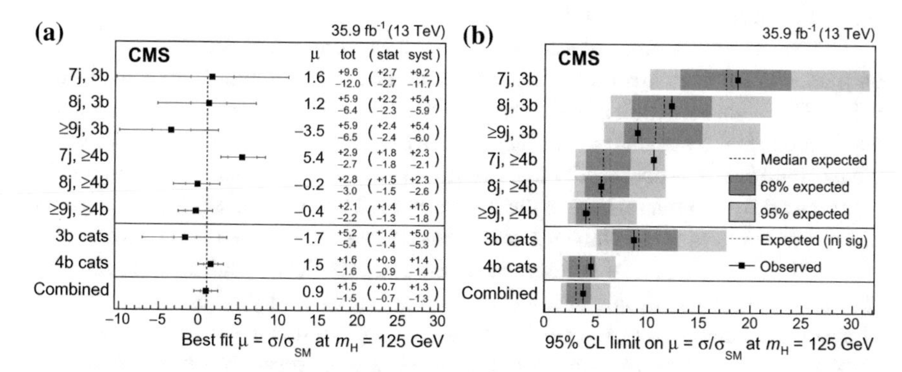

Fig. 7.7 (**a**) Best fit values of the signal strength modifiers μ with their 68% CL intervals, split into the statistical and systematic components. (**b**) Median expected and observed 95% CL upper limits on μ. The expected limits are displayed together with their 68 and 95% CL intervals, and with the expectation for an injected signal of $\mu = 1$ (inj sig)

Since the measured signal strength is also compatible with the background-only hypothesis, an exclusion limit at the 95% CL can be set using the modified frequentist CL_s method and the asymptotic approximation described in Sect. 7.1. Combining all categories, the observed and expected upper limits are $\mu < 3.8$ and $\mu < 3.1$, respectively. The expected upper limit under the signal+background hypothesis is $\mu < 3.9$, which is consistent with the observed limit for $\hat{\mu} = 0.9$. The observed and expected upper limits in each category as well as for the combined fit in all categories, are listed in Table 7.3 and displayed in Fig. 7.7b.

From Table 7.3 it is seen that a large signal strength is fitted in the 7j, ≥4b category. This large value is attributed to the excess observed in the last bin of the

MEM discriminant in Fig. 7.4b. Since the relative contribution of each category in a combined fit is given by the uncertainty on $\hat{\mu}$, this large excess is diluted by the negative $\hat{\mu}$ values obtained in the 8j, ≥4b and ≥9j, ≥4b categories to give a best fit μ value of 1.4 in the combined 4b category fit. The result in the 3b category fit is dominated by the large negative μ value fitted in the 9j, 3b category. The value of $\hat{\mu}$ in the combined fit is then a weighted average of the values obtained from the fit to the 3b and 4b categories, where lower uncertainties translate into higher weights. Given this, the best fit value of μ from the combined fit is positive thanks to the large excess observed in the 7j, ≥4b category.

A similar story can be told for the observed exclusion limits. The higher observed limit relative to the expected limit in the combined fit is due to the large limit observed in the 7j, ≥4b category, which is in turn caused by the excess in the 10th bin of its MEM discriminant. The sensitivity of the search is given by the expected upper limit, which is driven by the 4b categories, particularly the ≥9j, ≥4b category. The expected upper limit of 3.1 indicates that the background prediction is able to exclude a signal with strength $\mu = 3.1$ at the 95% CL.[2] The benefit of including the signal-poor 3b categories is seen by the improvement of the expected upper limit and reduction in the uncertainty on $\hat{\mu}$ in the combined fit compared to the fit in the 4b categories. It is small, but an improvement nonetheless.

The overall significance of the signal is small as indicated by the large uncertainty on $\hat{\mu}$ and expected exclusion limit greater than unity. Nevertheless, since the best fit value of the signal strength is positive, its significance can be calculated. The observed significance is 0.6 standard deviations, with a corresponding p-value of 0.26, which means there is a 26% probability that the observed value of the test statistic or greater is produced from a fluctuation of the background. The expected significance for the SM signal strength is 0.7 standard deviations, corresponding to a p-value of 0.25.

Given the dominance of systematic uncertainties in these results, future searches for $\mathrm{t\bar{t}H}$ in the fully hadronic decay channel must significantly reduce many sources of systematic uncertainty in order to achieve a 5σ significance – more data, while certainly useful, is simply not enough to reach a discovery. Future efforts to constrain many important sources of systematic uncertainty and improve the background rejection are discussed in the Outlook. Furthermore, this analysis can make a significant contribution to the overall $\mathrm{t\bar{t}H}$ ($\mathrm{H} \rightarrow \mathrm{b\bar{b}}$) search, which could be the difference between a future discovery in this channel and systematic limitation.

7.2.4 *Comparison to Previous Results*

Since this is the first search for $\mathrm{t\bar{t}H}$ production in the fully hadronic decay channel at 13 TeV, no direct comparisons can be made, and this result sets the benchmark in this channel. The only other fully hadronic $\mathrm{t\bar{t}H}$ search has been performed by ATLAS

[2] A search becomes sensitive to a signal when the expected exclusion limit lies below $\mu = 1$. In such cases, the significance would be above 2σ, with a p-value of less than 0.05.

with $20.3\,\text{fb}^{-1}$ of data at $\sqrt{s} = 8\,\text{TeV}$ [5]. In that search, observed and expected upper limits were set at the 95% CL of $\mu < 6.4$ and $\mu < 5.4$, respectively, and a best-fit value for the signal strength relative to the SM expectation of $\mu = 1.6 \pm 2.6$ was obtained. Since the present result is better than the previous published result, it represents the best ever measurement of $\text{t}\bar{\text{t}}\text{H}$ production, in which the Higgs boson decays to $\text{b}\bar{\text{b}}$ and the top quarks decay to hadrons.

To compare the two available results on an equal footing, an adjustment can be made for the centre-of-mass energy. Increasing $\sqrt{s}$ from 8 TeV to 13 TeV results in an increase of the production cross section for most processes. Comparing the yields in the 7j, 3b category of the two analyses and adjusting for the integrated luminosity, the following factors are obtained, which approximately represent the increases in the cross section for the most relevant processes:

- $\text{t}\bar{\text{t}}\text{H}$: 3.4
- $\text{t}\bar{\text{t}}$ + jets: 3.2
- QCD multijet: 2.2

These can be used to scale the signal and background contributions in the most sensitive category from 8 TeV to 13 TeV. With the additional scaling of the luminosity, an estimate can be derived for the expected limits, assuming that the relative contribution of systematic uncertainties remains the same. With this method, the expected upper limit obtained by ATLAS at 8 TeV roughly corresponds to an upper limit of $\mu < 1.9$ at 13 TeV with $35.9\,\text{fb}^{-1}$ of data. However, there are a few caveats to the simple extrapolation to a higher centre-of-mass energy and instantaneous luminosity: systematic uncertainties are expected to have a more dominant impact; the rejection of QCD multijet events becomes more difficult; and the signal efficiency decreases.

7.3 Combination with Other Analyses

As discussed in Chap. 1 and Sect. 2.4.3, the search for $\text{t}\bar{\text{t}}\text{H}$ production at CMS is conducted in various decay channels of the Higgs boson, and also the top quarks. Specifically, the individual searches considered are:

- $\text{t}\bar{\text{t}}\text{H}$ ($\text{H} \to \text{ZZ}$). Considered as part of the inclusive $\text{H} \to \text{ZZ}$ search [6].
- $\text{t}\bar{\text{t}}\text{H}$ ($\text{H} \to \gamma\gamma$). Considered as part of the inclusive $\text{H} \to \gamma\gamma$ search [7].
- $\text{t}\bar{\text{t}}\text{H}$ ($\text{H} \to \text{WW}$). Commonly referred to as $\text{t}\bar{\text{t}}\text{H}$ multilepton [8].
- $\text{t}\bar{\text{t}}\text{H}$ ($\text{H} \to \tau\tau$) [9]. Closely related to and combined with the multilepton search [10].
- Leptonic $\text{t}\bar{\text{t}}\text{H}$ ($\text{H} \to \text{b}\bar{\text{b}}$). Includes the single-lepton and di-lepton decay channels of the top quark pair [11].
- Hadronic $\text{t}\bar{\text{t}}\text{H}$ ($\text{H} \to \text{b}\bar{\text{b}}$). This analysis [12].

A combination of all the above $\text{t}\bar{\text{t}}\text{H}$ searches and the combined $\text{t}\bar{\text{t}}\text{H}$ results from Run 1 [13] has been performed [14], which results in the first ever observation of $\text{t}\bar{\text{t}}\text{H}$ production. The resulting observed and expected significances are 5.2 and 4.2 standard deviations, respectively, and the corresponding best fit value of the signal

Table 7.4 Best fit values of the signal-strength modifier μ in each Higgs boson decay channel and the combination, shown with their total uncertainties and the breakdown into the statistical (stat), experimental systematic (expt), background theory systematic (thbkg), and signal theory systematic (thsig) components

Channel	Best fit μ and uncertainty		
	$\hat{\mu}$	tot	(stat expt thbkg thsig)
$\mathrm{t\bar{t}H}$ ($\mathrm{H \to ZZ}$)	0.00	${}^{+1.30}_{-0.00}$	$\left({}^{+1.28}_{-0.00}\ {}^{+0.20}_{-0.00}\ {}^{+0.04}_{-0.00}\ {}^{+0.09}_{-0.00}\right)$
$\mathrm{t\bar{t}H}$ ($\mathrm{H} \to \gamma\gamma$)	2.27	${}^{+0.86}_{-0.74}$	$\left({}^{+0.80}_{-0.72}\ {}^{+0.15}_{-0.09}\ {}^{+0.02}_{-0.01}\ {}^{+0.29}_{-0.13}\right)$
$\mathrm{t\bar{t}H}$ ($\mathrm{H \to WW}$)	1.97	${}^{+0.71}_{-0.64}$	$\left({}^{+0.42}_{-0.41}\ {}^{+0.46}_{-0.42}\ {}^{+0.21}_{-0.21}\ {}^{+0.25}_{-0.12}\right)$
$\mathrm{t\bar{t}H}$ ($\mathrm{H} \to \tau\tau$)	0.28	${}^{+1.09}_{-0.96}$	$\left({}^{+0.86}_{-0.77}\ {}^{+0.64}_{-0.53}\ {}^{+0.10}_{-0.09}\ {}^{+0.20}_{-0.19}\right)$
$\mathrm{t\bar{t}H}$ ($\mathrm{H \to b\bar{b}}$)	0.82	${}^{+0.44}_{-0.42}$	$\left({}^{+0.23}_{-0.23}\ {}^{+0.24}_{-0.23}\ {}^{+0.27}_{-0.27}\ {}^{+0.11}_{-0.04}\right)$
13 TeV combined	1.14	${}^{+0.31}_{-0.27}$	$\left({}^{+0.17}_{-0.16}\ {}^{+0.17}_{-0.17}\ {}^{+0.13}_{-0.12}\ {}^{+0.14}_{-0.06}\right)$
Run 1 combined	2.59	${}^{+1.01}_{-0.88}$	$\left({}^{+0.54}_{-0.53}\ {}^{+0.53}_{-0.49}\ {}^{+0.55}_{-0.49}\ {}^{+0.37}_{-0.13}\right)$
$\mathrm{t\bar{t}H}$ combined	1.26	${}^{+0.31}_{-0.26}$	$\left({}^{+0.16}_{-0.16}\ {}^{+0.17}_{-0.15}\ {}^{+0.14}_{-0.13}\ {}^{+0.15}_{-0.07}\right)$

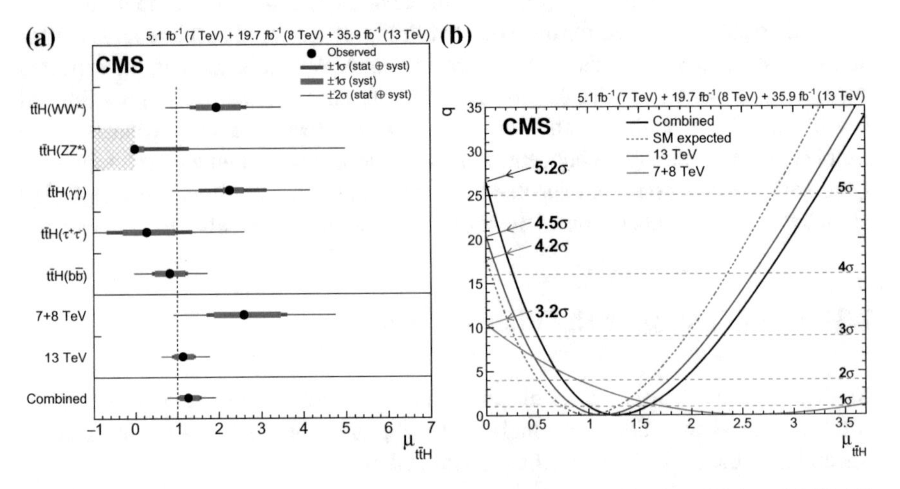

Fig. 7.8 (**a**) Best fit values of the signal strength modifiers μ with their 68% and 95% CL intervals, split into the statistical and systematic components, by Higgs decay channel and combined. (**b**) Test statistic q as a function of $\mu_{\mathrm{t\bar{t}H}}$ for the various combinations and that expected for the overall combination. The horizontal dashed lines indicate the significance for a given value of μ, and the usual discovery significance is indicated at the background-only hypothesis, $\mu = 0$

strength is $\mu = 1.26^{+0.31}_{-0.26}$, which is in agreement with the SM expectation. The results are listed with a breakdown by decay channel in Table 7.4 and shown in Fig. 7.8.

I have performed the overall combinations and the $\mathrm{t\bar{t}H}$ ($\mathrm{H \to b\bar{b}}$) combination with and without the fully hadronic $\mathrm{t\bar{t}H}$ analysis, with the results summarised in Table 7.5. Excluding the fully hadronic analysis, the overall best fit signal strength is $\hat{\mu} = 1.24^{+0.32}_{-0.27}$ and the observed (expected) significance is 4.98 (4.14), indicating

Table 7.5 Best fit values of the signal-strength modifier μ in each combination, and the observed and expected significances, excluding the contribution of the fully hadronic (FH) analysis. The results including the FH analysis are also shown for comparison

Channel	Excluding FH			Including FH		
	Best fit μ	Significance		Best fit μ	Significance	
		Obs.	Exp.		Obs.	Exp.
$t\bar{t}H\ (H \rightarrow b\bar{b})$	$0.72^{+0.45}_{-0.45}$	1.55	2.16	$0.82^{+0.44}_{-0.42}$	2.01	2.33
13 TeV combined	$1.11^{+0.31}_{-0.28}$	4.33	3.98	$1.14^{+0.31}_{-0.27}$	4.55	4.09
$t\bar{t}H$ combined	$1.24^{+0.32}_{-0.27}$	4.98	4.14	$1.26^{+0.31}_{-0.26}$	5.21	4.24

that this analysis indeed plays a crucial role in the observation of $t\bar{t}H$ production. Furthermore, the fully hadronic channel makes a significant contribution to the $t\bar{t}H\ (H \rightarrow b\bar{b})$ result, increasing the expected sensitivity by around 8%. The contribution is larger than that implied by a simple sum in quadrature, thanks to the correlations between systematic uncertainties, many of which are constrained by the leptonic $t\bar{t}H\ (H \rightarrow b\bar{b})$ analysis. In particular, b-tagging uncertainties and $t\bar{t}$ + heavy-flavour normalisation uncertainties are significantly constrained, thus enhancing the sensitivity of the fully hadronic search.

References

1. The ATLAS Collaboration, The CMS Collaboration, The LHC Higgs Combination Group (2011) Procedure for the LHC Higgs boson search combination in Summer 2011. ATL-PHYS-PUB-2011-11, CMS-NOTE-2011-005 http://cds.cern.ch/record/1379837
2. Read AL (2002) Presentation of search results: the CL_s technique. J Phys G 28:2693. https://doi.org/10.1088/0954-3899/28/10/313
3. Junk T (1999) Confidence level computation for combining searches with small statistics. Nucl Instrum Meth A 434:435. https://doi.org/10.1016/S0168-9002(99)00498-2. arXiv:hep-ex/9902006
4. Cowan G, Cranmer K, Gross E, Vitells O (2011) Asymptotic formulae for likelihood-based tests of new physics. Eur Phys J C 71:1554. https://doi.org/10.1140/epjc/s10052-013-2501-z, arXiv:1007.1727. [Erratum: Eur Phys J C 73:2501 (2013)]
5. ATLAS Collaboration (2016) Search for the standard model Higgs boson decaying into $b\bar{b}$ produced in association with top quarks decaying hadronically in pp collisions at $\sqrt{s} = 8$ TeV with the ATLAS detector. JHEP 05:160, https://doi.org/10.1007/JHEP05(2016)160, arXiv:1604.03812
6. CMS Collaboration (2017) Measurements of properties of the Higgs boson decaying into the four-lepton final state in pp collisions at sqrt(s) = 13 TeV. JHEP 11:047. https://doi.org/10.1007/JHEP11(2017)047, arXiv:1706.09936
7. CMS Collaboration (2018) Measurements of Higgs boson properties in the diphoton decay channel in proton-proton collisions at $\sqrt{s} = 13$ TeV. JHEP 11:185. https://doi.org/10.1007/JHEP11(2018)85, arXiv:1804.02716
8. CMS Collaboration (2017) Search for Higgs boson production in association with top quarks in multilepton final states at $\sqrt{s} = 13$ TeV. CMS-PAS-HIG-17-004 https://cds.cern.ch/record/2256103

9. CMS Collaboration (2017) Search for the associated production of a Higgs boson with a top quark pair in final states with a τ lepton at $\sqrt{s} = 13$ TeV. CMS-PAS-HIG-17-003 https://cds.cern.ch/record/2257067
10. CMS Collaboration (2018) Evidence for associated production of a Higgs boson with a top quark pair in final states with electrons, muons, and hadronically decaying τ leptons at $\sqrt{s} = 13$ TeV. JHEP 08:066. https://doi.org/10.1007/JHEP08(2018)066, arXiv:1803.05485
11. CMS Collaboration (2019) Search for $\mathrm{t\bar{t}H}$ production in the $\mathrm{H} \rightarrow \mathrm{b\bar{b}}$ decay channel with leptonic $\mathrm{t\bar{t}}$ decays in proton-proton collisions at $\sqrt{s} = 13$ TeV. JHEP 03:026. https://doi.org/10.1007/JHEP03(2019)026, arXiv:1804.03682
12. CMS Collaboration (2018) Search for $\mathrm{t\bar{t}H}$ production in the all-jet final state in proton-proton collisions at $\sqrt{s} = 13$ TeV. JHEP 06:101. https://doi.org/10.1007/JHEP06(2018)101, arXiv:1803.06986
13. CMS Collaboration (2014) Search for the associated production of the Higgs boson with a top-quark pair. JHEP 09:087. https://doi.org/10.1007/JHEP10(2014)106, arXiv:1408.1682. [Erratum: JHEP 10:106 (2014)]
14. CMS Collaboration (2018) Observation of $\mathrm{t\bar{t}H}$ production. Phys Rev Lett 120:23, 231801 (2018). https://doi.org/10.1103/PhysRevLett.120.231801, arXiv:1804.02610

Chapter 8
Conclusions

The search for the standard model (SM) Higgs boson produced in association with top quarks in the fully hadronic final state, using 35.9 fb^{-1} of proton-proton collision data collected by the CMS experiment at 13 TeV, has been presented. The $\mathrm{t\bar{t}H}$ production mode provides access to a direct measurement of the top-Higgs Yukawa coupling, which would be a strong test of the SM. An observation of $\mathrm{t\bar{t}H}$ production has not yet been made, with the strongest evidence to date falling short of the 5σ significance required for a discovery. Nevertheless, the significance is increasing and the results from this search will bring us closer to the discovery threshold.

The search selects events online using dedicated all-jet triggers, and offline using kinematic requirements on jets as well as b-tagging and quark-gluon discrimination criteria. Specifically, events for analysis are required to have no muons or electrons, at least 7 jets, of which 3 or more are b tagged, and untagged jets that are more quark-like than gluon-like. Six orthogonal categories are formed based on jet and b-tag multiplicity and a matrix element method (MEM) is used to assign a signal and background probability density to each event.

The MEM uses the full event information to reconstruct the phase space of the quark-level final state, consisting of 4 b jets and 4 light-flavour jets. It then sums over all possible combinations of jet-quark associations to reconstruct the tree level $\mathrm{t\bar{t}H}$ and $\mathrm{t\bar{t}+b\bar{b}}$ processes. It integrates over poorly measured or missing variables in the calculation of a probability density for each process, which is based on the leading order production cross section. The two probability densities are combined in a single likelihood discriminant, by which the signal is extracted.

A binned maximum likelihood fit is performed to the MEM discriminant in all six categories combined, where the parameter of interest is the $\mathrm{t\bar{t}H}$ signal strength, μ, defined as the ratio of the measured $\mathrm{t\bar{t}H}$ cross section to that predicted by the SM. The resulting best-fitted value of the signal strength is $\hat{\mu} = 0.9 \pm 1.5$, which is compatible with the SM expectation. The observed and expected significance of the signal are 0.6 and 0.7 standard deviations, respectively. Since the data are also

D. Salerno, *The Higgs Boson Produced With Top Quarks in Fully Hadronic Signatures*, Springer Theses, https://doi.org/10.1007/978-3-030-31257-2_8

compatible with the background-only hypothesis, observed and expected exclusion limits of $\mu < 3.8$ and $\mu < 3.1$, respectively, are obtained at the 95% confidence level.

From a combination of this search with other $t\bar{t}H$ searches at CMS, observed and expected significances of 5.2 and 4.2 standard deviations, respectively, are obtained. The corresponding best fit value of the signal strength is $\mu = 1.26^{+0.31}_{-0.26}$, which is in agreement with the SM expectation. The results of this analysis improve the observed significance of the combined $t\bar{t}H$ search by around 5%, and bring it over the 5σ threshold required to claim an observation.

Outlook

The future of this analysis appears very promising. More data will lead to more stringent limits and a better significance. An estimate for the sensitivity of the analysis at higher luminosities and the same centre-of-mass energy, assuming the relative contribution of systematic uncertainties remains constant, can be obtained by a simple scaling of the present results based on the signal and background contributions in the most sensitive category. With this method, the following 95% confidence level median expected limits and significances for different amounts of integrated luminosity are projected:

	Integrated luminosity (fb^{-1})				
	100	200	300	500	1000
Expected limit	1.9	1.3	1.1	0.8	0.6
Significance	1.1	1.5	1.9	2.4	3.4

These projections and the scaling of systematic uncertainties implied within, are expected to be achieved through both enhanced background rejection and a reduction of the systematic uncertainties.

There are several ideas to improve the background rejection, particularly of the QCD multijet process, which include: a jet-based use of the quark-gluon likelihood discriminator, in which the information is used to determine if individual jets are selected or ignored; the use of multivariate techniques such as boosted decision trees or deep neural networks to further discriminate against the multijet background, either as a selection requirement or as part of the final discriminant; and a representative matrix element process for QCD multijet, which would result in a third MEM event probability density to specifically target this background.

Many systematic uncertainties are expected to be reduced through better measurements and/or calculations: the b-tagging scale factors for charm-flavour jets will be accurately measured in data; the $t\bar{t}$ + heavy-flavour cross sections are expected to be measured with greater precision and higher order calculations are expected to constraint their uncertainty; techniques to constrain the QCD multijet normalisation will be investigated, including the possibility of using control regions in the final fit; a better modelling of the multijet background will be attempted through more complex data-driven estimation methods; and a better handle on the quark-gluon likelihood reweighting will be obtained through the use of larger control samples.

Overall, the future developments of the analysis are expected to achieve or even slightly improve on the projected results. Together with the leptonic $\mathrm{t\bar{t}H}$ ($\mathrm{H} \rightarrow \mathrm{b\bar{b}}$) search, discovery potential is expected to be reached within the next five years. Combining with all $\mathrm{t\bar{t}H}$ searches at CMS, a 5σ significance has already been observed, paving the way for higher precision direct measurements of the top quark Yukawa coupling. Future direct measurements will also improve the precision in κ_{t}, which is currently measured by CMS to be around 12%.

Appendix A
MEM Studies

In this appendix additional studies leading up to the final version of the MEM discussed in Chap. 5 are provided.

A.1 QCD Matrix Elements

As mentioned in Sect. 5.2.1, I investigated the use of a representative matrix element process for the QCD multijet background. Four different $2 \rightarrow N$ processes were considered as summarised in Table 5.1 and repeated here:

- gg $\rightarrow$ gg: where each pair of b-tagged jets is assumed to come from a gluon decay, while the untagged jets are ignored (2jQCD).
- gg $\rightarrow$ ggg: where the reconstructed top quarks and Higgs boson are assumed to be gluons (3jQCD).
- gg $\rightarrow$ gggg: where each pair of b-tagged and untagged jets is assumed to come from a gluon (4jQCD).
- gg $\rightarrow$ $\mathrm{b\bar{b}b\bar{b}}$: where each b-tagged jet is assumed to come from the LO matrix element process and the untagged jets are ignored (4bQCD).

The signal vs. background efficiencies for simulated QCD multijet events with eight jets and four or more b tags are compared for each of the above matrix elements as well as the default $\mathrm{t\bar{t}+b\bar{b}}$ matrix element (ttbb) in Fig. A.1. As can be seen, the performance of the default matrix element is better than all representative matrix elements considered. The three njQCD processes with $n = 2, 3, 4$ were studied using only partial statistics, which, when combined with the low number of simulated multijet events passing the selection, results in jagged ROC curves.

D. Salerno, *The Higgs Boson Produced With Top Quarks in Fully Hadronic Signatures*, Springer Theses, https://doi.org/10.1007/978-3-030-31257-2

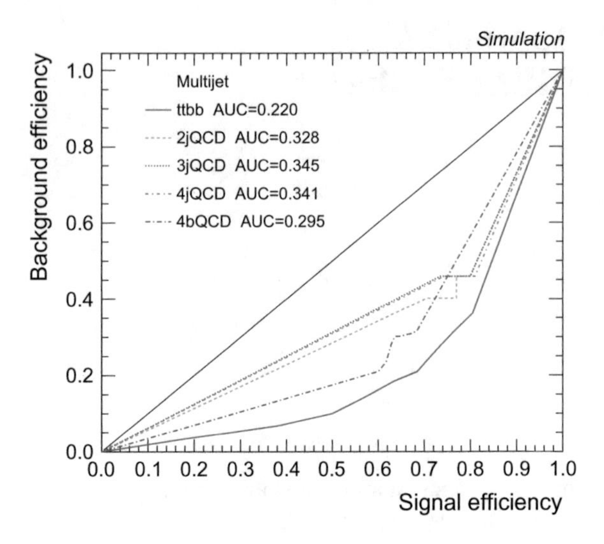

Fig. A.1 ROC curves for different background matrix element processes in simulated $t\bar{t}H$ and multijet events with 8 jets and ≥4 b tags

A.2 Lost Quark Hypotheses

As discussed in Sect. 5.4.1, I studied several different hypotheses regarding lost quarks in the matrix element calculation. The different hypotheses are summarised in Table 5.2 and the performance of some of these in terms of ROC curves is shown in Fig. A.2. In each case, the maximum reconstructed hypothesis is shown together with some lost quark hypothesis. In the case of a lost quark, an integration is performed over its direction and where necessary its energy. The different hypotheses shown in the figure are as follows:

- int. 1q: a single quark from a W boson decay is assumed lost and its direction is integrated over.
- int. 1b: a single bottom quark from a top quark decay is assumed lost and its direction is integrated over.
- int. 2q: a single quark from each W boson decay is assumed lost and their directions are integrated over.
- int. 1W: both quarks from a W boson decay are assumed lost. The energy of one quark and the directions of both are integrated over.
- int. 1q,1b: a single quark from a W boson decay and a bottom quark from a top quark decay are assumed lost and their directions are integrated over.

In most cases the maximally reconstructed hypothesis performs best, with the exception of the 8 jet, ≥4 b tag and ≥9 jet, ≥4 b tag events. In the case of 8 jets, both the 1q and 2q lost quark hypotheses perform best, and the single lost quark hypothesis is adopted due to its faster calculation. In the case of ≥9 jets, the ROC curves shown only consider the first 8 jets. This was later improved to permute over the 9th jet then

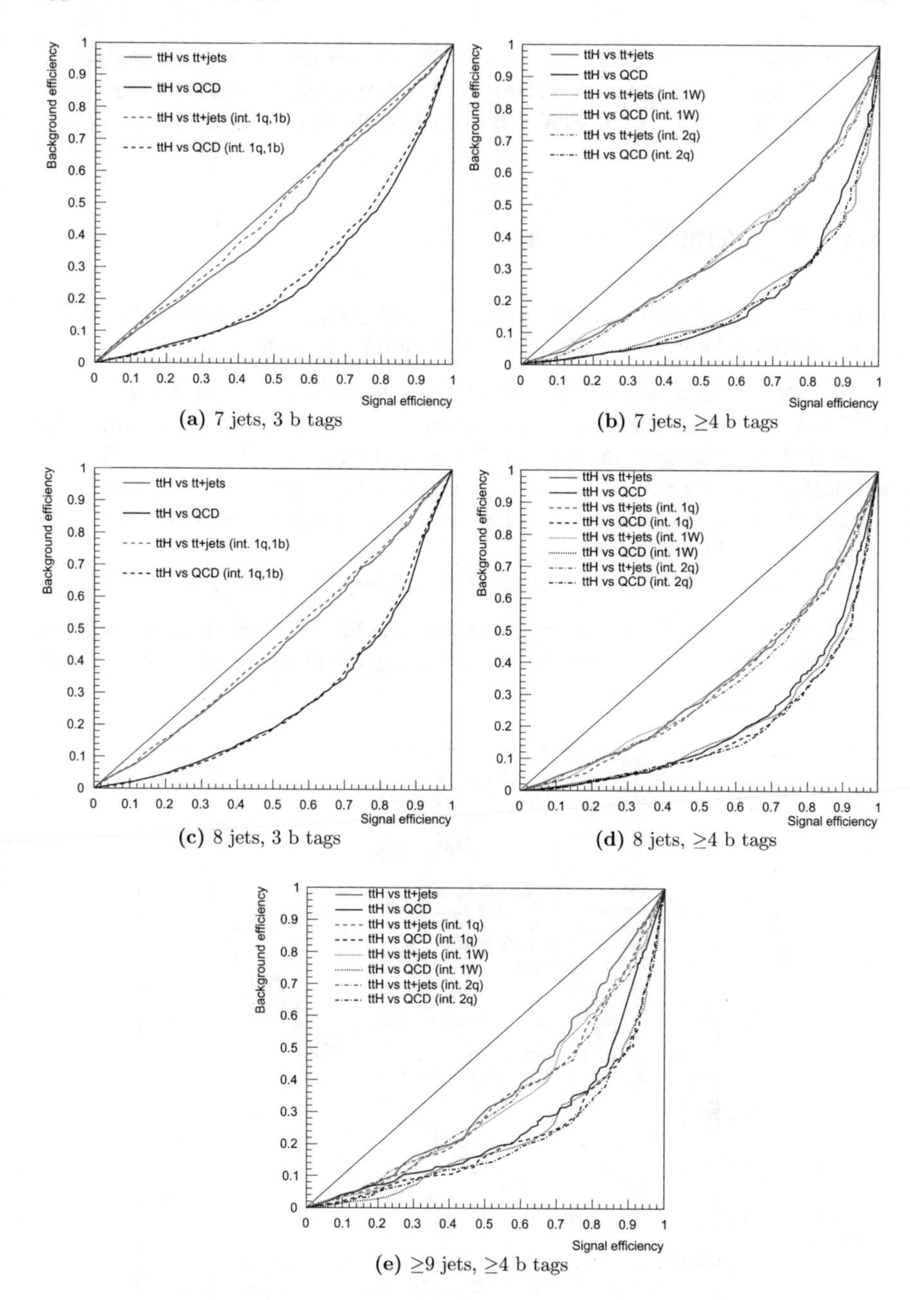

Fig. A.2 ROC curves for different lost quark hypotheses in simulated $t\bar{t}H$, $t\bar{t}$ + jets and QCD multijet events. The assumed hypothesis is shown in parenthesis in the legend, where the lost quarks are integrated over as described in the text the performance of the fully reconstructed hypothesis improves. Nevertheless, the 1q and 2q hypotheses still perform better, but unfortunately their computational time is prohibitive. The final choices of lost quark hypothesis are shown in Table 5.6.

A.3 B-Tagging Algorithms

I also studied the performance of different b-tagging algorithms in use at CMS. The four algorithms I considered are described in Ref. [1] and summarised below:

- CSV: the combined secondary vertex algorithm described in Sect. 4.3.6.
- CMVA: the combined multivariate analysis tagger uses six b jet identification discriminators as input variables, including two variants of the CSV algorithm.
- DeepCSV: a new version of the CSV tagger using a deep neural network with more hidden layers, more nodes per layer, and a simultaneous training in all vertex categories and for all jet flavours.
- DeepCMVA: similar to the CMVA but using DeepCSV inputs instead of CSV.

In addition to the medium working point selection for b-tagged jets (nBCSVM and nBCMVAM), I also investigated the use of a b-tagging likelihood ratio (blr), similar

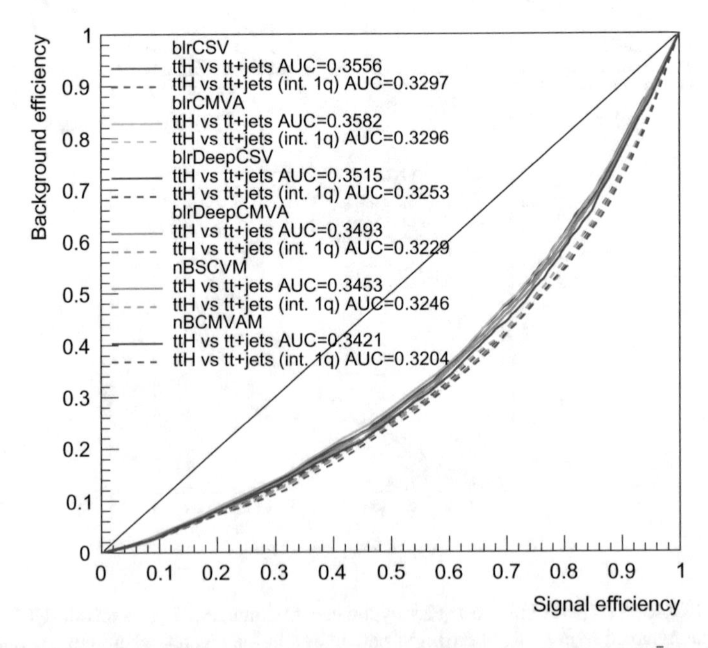

Fig. A.3 ROC curves for different b-tagging algorithms and selections in simulated $t\bar{t}H$ and $t\bar{t}$ + jets events with 8 jets and ≥4 b tags. Blr indicates that the selection and identification of b-tagged jets is performed via a b-tagging likelihood ratio, while nB indicates that the identification is made by a simple counting of jets passing the medium b-tagging working point to the QGLR and described in Ref. [2], for the selection and identification of b jets. The performance of the different b-tagging algorithms and the b-tag selections in simulated $t\bar{t}H$ and $t\bar{t}$ + jets events with eight jets and four or more b tags can be seen in the ROC curves of Fig. A.3.

As can be seen in Fig. A.3, the performance of the different algorithms and selections is similar. Given this, the choice to use the CSV algorithm and a simple counting of b-tagged jets was made based on ease of computation and the availability of scale factors and associated systematic uncertainties.

Appendix B
Example Statistical Calculation

In this appendix, I work through the calculation of the results for one category, namely the ≥9j, ≥4b category. To simplify the computations, systematic uncertainties are excluded, however the extension to one nuisance parameter is straightforward, and thus so is the extension to N nuisance parameters.

The bin-by-bin event yields in the MEM discriminant for the signal, background and data, taken from Fig. 6.17f, are listed in Table B.1.

B.1 Maximum Likelihood

Finding the maximum of the likelihood given in Eq. 7.3, is equivalent to finding the maximum of its logarithm:

$$\ln(\mathcal{L}) = \sum_{i=1}^{10} \Big[n_i \ln(\mu s_i + b_i) - \ln(n_i!) - (\mu s_i + b_i) + \ln p(\tilde{\theta}|\theta) \Big] . \tag{B.1}$$

Without systematic uncertainties the last term vanishes. Furthermore the term independent of μ can be ignored. The equation to maximise then becomes:

$$\ln(L) = \sum_{i=1}^{10} \Big[n_i \ln(\mu s_i + b_i) - (\mu s_i + b_i) \Big], \tag{B.2}$$

which is shown as a function of μ in Fig. B.1. The maximum likelihood estimator of μ is found to be $\hat{\mu} = 0.27$.

D. Salerno, *The Higgs Boson Produced With Top Quarks in Fully Hadronic Signatures*, Springer Theses, https://doi.org/10.1007/978-3-030-31257-2

Table B.1 Individual bin yields of the MEM discriminant in the ≥9j, ≥4b category, for the signal, total background and data

Bin	1	2	3	4	5	6	7	8	9	10	Total
Signal	10.5	2.3	2.2	2.2	2.5	3.0	3.9	6.1	9.9	7.8	50.3
Background	1 690	229	188	170	178	177	200	252	297	134	3 516
Data	1 727	260	195	164	171	178	211	239	274	147	3 566

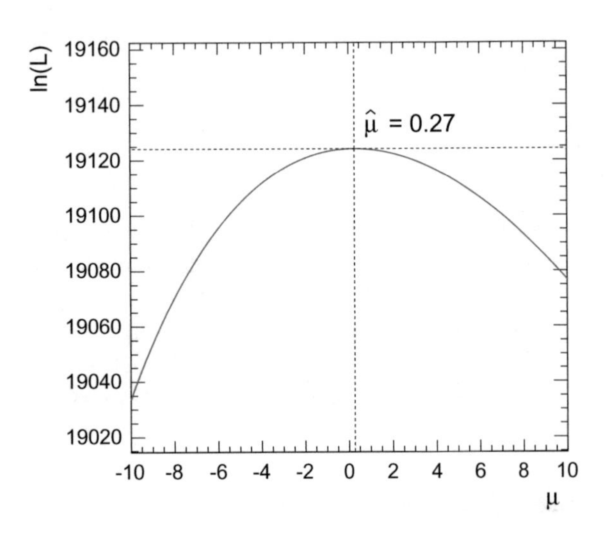

Fig. B.1 Log-likelihood given by Eq. (B.2) as a function of μ for the ≥9j, ≥4b category

B.2 Upper Limits

Finding the upper limits begins by determining the test statistic in Eq. (7.7). Given that there are no nuisance parameters in this example, and that the term with $n!$ is canceled out in the ratio, the test statistic becomes:

$$\tilde{q}_\mu = -2\Big[\ln(L|\mu) - \ln(L|\hat{\mu} = 0.27)\Big], \qquad \mu \geq 0.27. \tag{B.3}$$

The value of the test statistic as a function of μ is shown in Fig. B.2. For the signal+background hypothesis ($\mu = 1$), the observed value of the test statistic is $\tilde{q}_1^{\mathrm{obs}} = 0.63$.

To get the pdf of the test statistic under the asymptotic approximation, the variance of $\hat{\mu}$ must be estimated, using Eq. (7.10). This requires the calculation of the test static with the background-only Asimov data set. The resulting value of σ as a function of μ is shown in Fig. B.3.

The pdf and CDF of $\tilde{q}_\mu$ are given by Eqs. (7.11, 7.13) and (7.12, 7.14), and are shown in Fig. B.4 for $\mu = 1$ and 2. The p-values can be read directly from Fig. B.4b as $1 - F(\tilde{q}_\mu|\mu)$. The 95% CL upper limit is calculated to be 2.0 by adjusting the value of μ (s and b curves and $\tilde{q}_\mu^{\mathrm{obs}}$) until the following equation is satisfied:

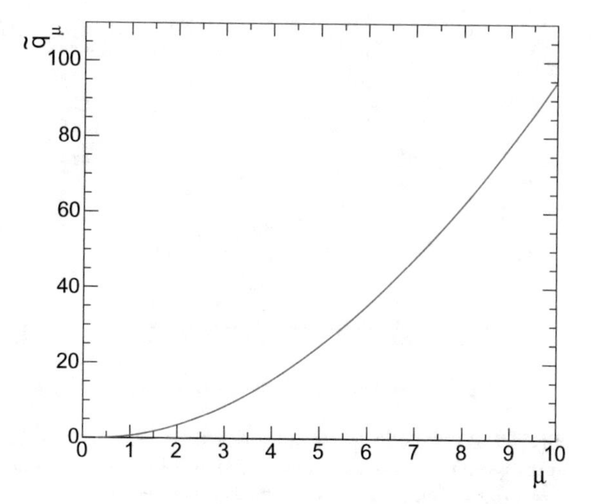

Fig. B.2 Test statistic given by Eq. (B.3) as a function of μ

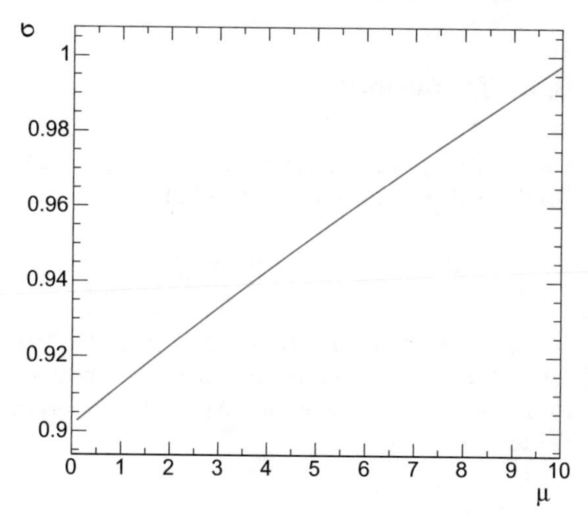

Fig. B.3 Estimated standard deviation of $\hat{\mu}$ with mean μ derived using the Asimov data set

$$0.05 = \frac{p_\mu(\tilde{q}_\mu^{\text{obs}})}{1 - p_b(\tilde{q}_\mu^{\text{obs}})} \quad \Rightarrow \quad \mu = 2.0. \tag{B.4}$$

The expected upper limits can be found by recalculating the maximum likelihood estimate $\hat{\mu}$ and the test statistic $\tilde{q}_\mu$ with the Asimov data set instead of the observed data, i.e. by replacing the last row in Table B.1 with the second-to-last row. The distributions in Fig. B.4 can then be remade to give the following 95% CL expected upper limits:

$$\text{Median: } 1.8, \qquad 68\% \text{ band: } (1.3, 2.5), \qquad 95\% \text{ band: } (1.0, 3.4). \tag{B.5}$$

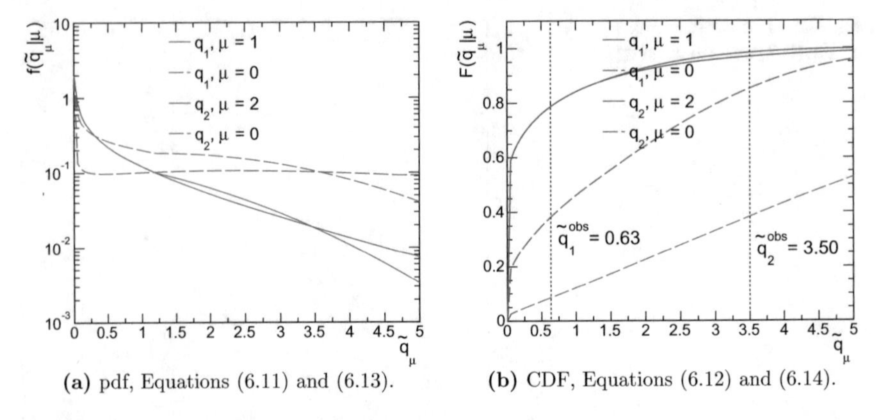

(a) pdf, Equations (6.11) and (6.13).

(b) CDF, Equations (6.12) and (6.14).

Fig. B.4 Asymptotic approximation of the distributions of the test statistic under the hypotheses of background-only (dashed line) and signal+background (solid line) for testing signal strengths of $\mu = 1$ (upper lines at $\tilde{q}_\mu \approx 2.5$) and 2 (lower lines)

B.3 Significance

Finally, the significance can be calculated starting with the test statistic in Eq. (7.18). For this case without nuisance parameters, q_0 is given by:

$$q_0 = 2\Big[\ln(L|\hat{\mu} = 0.27) - \ln(L|0)\Big], \tag{B.6}$$

which leads to an observed test statistic of $q_0^{\text{obs}} = 0.092$. The p-value can be directly calculated from Eqs. (7.20) and (7.21), which give $p_0 = 0.38$. The nature of $F(q_0|0)$ ensures that the Z-significance is equal to the square root of the observed value of the test statistic, i.e. 0.30σ.

References

1. CMS Collaboration (2018) Identification of heavy-flavour jets with the CMS detector in pp collisions at 13 TeV. JINST 13(05):P05011. http://dx.doi.org/10.1088/1748-0221/13/05/P05011, arXiv:1712.07158
2. CMS Collaboration (2015) Search for a standard model higgs boson produced in association with a top-quark pair and decaying to bottom quarks using a matrix element method. Eur Phys J C 75:251. http://dx.doi.org/10.1140/epjc/s10052-015-3454-1, arXiv:1502.02485

D. Salerno, *The Higgs Boson Produced With Top Quarks in Fully Hadronic Signatures*, Springer Theses, https://doi.org/10.1007/978-3-030-31257-2

Printed in the United States
By Bookmasters